ERRATA.

The word "treasurers," on page 11, in the 6th line, should read "treasures."

The word "deposits," on page 14, in the 18th line, should read "deposit."

In the head line, or caption of Chapter XXVI, on page 193, the word "(Jicarilla,)" should read "(Jarilla.)"

The word "ehloride," on page 284, in the 21st line, should read "chloride."

On page 301, under the caption of San Antonio Springs, an "a" is omitted in the word "and", at the beginning of the second line, which should be supplied.

On page 304, at the explanation in small print immediatly after the analyses of the Coyote (Chaves) Spring and Topham Artesian (well) spring, the word "analysed" should read "analyzed."

On page 308, in the last word in the explanation in small print that is immediately below the analysis of the Spring on Penasco, Chaves County, should read "aperient," instead of "aparient."

In biographical sketch of Charles A. Spiess, in appendix, fourteenth line from top, read "1894" instead of "1895".

In biographical sketch of Herbert J. Hagerman, in appendix, in eighth line, read "1894" instead of "1904".

WITH THE COMPLIMENTS OF

THE BUREAU OF IMMIGRATION
OF THE
TERRITORY OF NEW MEXICO,

ESPIRITU SANTO LAKE, IN THE SANTA FE RANGE.

NEW MEXICO

MINES AND MINERALS.

WORLD'S FAIR EDITION, 1904.

Being an Epitome of the Early Mining History and Resources
of New Mexican Mines, in the Various Districts, Down
to the Present Time. Geology of the Ore Deposits,
Complete Census of Minerals, Mineral and Irrigation Waters, Table of Altitudes and
Other General Information.

ILLUSTRATED.

" Surely there is a vein for the silver, and a place for the gold
where they fine it."—Job 28:1.

BY

FAYETTE ALEXANDER JONES, C. E., E. M., LL.D.,

Member of the Territorial Board of the Louisiana Purchase
Exposition Managers and Director of the Mineral Exhibit.

SANTA FE, N. M.:
THE NEW MEXICAN PRINTING COMPANY,
1904.

TO

THE MEMORY OF THE 19th CENTURY PROSPECTORS

OF

NEW MEXICO,

This Volume is Most Sincerely Dedicated

by the

Territorial Board

of the

Louisiana Purchase Exposition Managers.

The passing of the " old time " prospector forever removes from
the American people one of the most unique characters
in the history of the Republic.

.

LIST OF ILLUSTRATIONS.

CONTENTS.

2   CONTENTS.

PREFACE.

Through the medium of the Territorial Board of the Louisiana Purchase Exposition Managers of New Mexico, the publication of this volume on New Mexico Mines and Minerals was made possible.

The Board of Managers fully recognized, in the beginning, the importance of presenting in a proper and concise manner such matters of interest, at the Great Exposition in the City of St. Louis, concerning New Mexico, as would redound to the general welfare of the commonwealth and at the same time tend to reflect credit on itself in the performance of its manifold duties, has in accordance with such views, sanctioned the publication of this brief historical Memoir descriptive of the Mines and Minerals of New Mexico. It is sincerely hoped by the Commission that the matter herein contained will be cordially received by a generous public and may prove both interesting and valuable to all from a scientific, as well as from a historical standpoint. Technical terms and phrases have been avoided as much as was consistent in elucidating intelligently the various topics discussed.

Should any criticism be offered as to the manner, character and style of presenting the subjects contained in "New Mexico Mines and Minerals," the author desires the other members of the board exonerated from such rebuke, and will himself shoulder the full responsibility; since the whole volume was entirely written by him and was left in his hands, absolutely.

Moreover, the author wishes to say that he gratuitously prepared the whole of the manuscript and bore the entire burden and expense of collation, correspondence and stenography.

Every effort was exerted to secure and include only that which is authentic, and whenever possible, verified by living representatives of the "early days," who were on the ground in person. It should be observed then, that the source of much information thus obtained, was not through mere hearsay or by second-hand evidence. Such evidence as was gotten can be gathered at this time with more accuracy than would be possible so to do a decade hence.

The fact should not be overlooked, that the surviving prospectors of the "early days" are few in number, when compared to the vast army that have followed the inexorable command of the Silent Captain to explore the untrodden regions of the Great Unknown.

The passing of the "old time" prospector, forever removes from the American people one of the most unique characters in the history of the Republic.

The author is indebted to numerous persons throughout the Territory for valuable aid and data furnished in preparing this volume; among those deserving special mention for their kindness are:

James Lynch, J. M. Webster, Prof. J. S. Macgregor, J. P. Rinker, John Y. Hewitt, Capt. M. Cooney, Charles R. Smith, F. B. Schermerhorn, Major W. H. H. Llewellyn, Col. G. W. Prichard, Col. A. W. Harris, H. Lesdos, Hon. L. B. Prince, B. D. Wilson, J. Van Houten, Hon. J. M. Abbott, Thos. A. Lister, G. L. Brooks, J. C. Plemmons, Dr. Charles R. Keyes, Hon. Seaman Field, David Stitzel, Dr. M. M. Crocker, J. S. Hutchason, M. W. Porterfield, E. L. Smart, D. S. Miller, J. G. Schumann, W. J. Weatherby, David Egelston, Ed. H. Smith, H. W. Russell, Jack Richardson, W. M. Woody, Hon. Antonio Joseph, A. R. Gibson, Arthur Seligman and Hon. W. S. Hopewell. The greater number of coal analyses were taken from the 1902-3 report of the Hon. Jo E. Sheridan, United States coal mine inspector.

To Prof. E. M. Skeats, of El Paso, many thanks are due for valuable service rendered in furnishing most of the analyses of river and well waters; especially, the Pecos Valley water analyses. And to M. E. Hickey, attorney at law, for a synopsis of the Mining Laws of New Mexico, the author is greatly indebted.

It is to be hoped that when the body corporate of the Territorial Board of the Louisiana Purchase Exposition Managers has been absolved by reason of the act which limits its existence, that it has left, at least, a foot print in the sands of the desert of time, faithfully embodied in this brief Memoir.

Very sincerely,

F. A. JONES,

Member of Board of Managers.

Albuquerque, N. M.,
July 4, 1904.

CHAPTER I.

GEOLOGY.

What is known concerning the geology of New Mexico at the present time is in the main fragmentary. New Mexico affords the largest unexplored field for geological research of any section in the Union.

With but few exceptions, not enough work has been done in any one part of the Territory, whereby the geological relations may be intelligently correlated with those in a different locality. Aside from the work of a few investigators, the general government has sorely neglected this region. Less than a dozen topographic sheets comprise the principal work done under the auspices of the United States Geological Survey; not a single complete quadrangle, of any particular locality, has been effected.

From what we have learned on this subject, we are very much indebted to Drs. Newberry and Hayden in their cursory, yet highly interesting investigations, in detached sections of the Territory; to Lieut. Wheeler in his geographical surveys west of the 100th meridian; to the valuable monograph by Captain Dutton, supplemented by Major Powell, on the plateau region in the vicinity of Mount Taylor; and later, by Dr. Herrick in his series of economic papers published in the bulletins of the University Geological Survey of New Mexico.

Valuable and comprehensive as these publications are, they serve only as a beginning, and are to be regarded as fragmentary in comparison with the great expanse of Territory that remains untouched by scientific investigation.

Nowhere in the world are the natural environments so favorable for the study of geological conditions, as exist in New Mexico. The exposure of the rock system is all that could be desired and may be read as the leaves of a book. The great tilted orogenic block composing the Sandia range, lying east of Albuquerque, is regarded as classic in this extraordinary type of mountain modeling. The bold escarp which faces the Rio Grande represents a perpendicular throw of fully five thousand feet.

The plateau region west of Mount Taylor and on toward the Grand Canyon is, perhaps, the most intensely interesting territory, where the effects of erosion may be observed, that can be found in the world. Here in this strange land of the Zuñis the geologist fancies he catches glimpses of the eternity

Fig. 1—"TOOTH OF TIME," near Acoma. Photographed by
W. M. Borrowdale, 1900.

of the whole past; yet, he is forced to admit that all of this high handed carving and sculpturing has taken place since the great Cretaceous rock system was laid down. Geologically speaking the horizon of his view borders on a comparatively recent period.

During Tertiary times New Mexico was a theater of volcanic activity; a greater portion of the eruptive flows took place during this geologic period. Most of the so called *mal pais* had their origin at, and just after the close of the Tertiary, lasting throughout the Pleistocene. Many of these lavas *(mal pais)* are so recent that it is thought by some that the early Spanish explorers may have witnessed some of the lingering volcanic outbreaks. It has been claimed that the destruction of the Gran Quivira and some other Pueblos was due to causes of this nature. In the writer's opinion, from personal observations at the Gran Quivira and elsewhere, such a conclusion cannot be verified and would seem untenable for lack of evidence. Centuries before the landing of Columbus, it is quite probable that the aborigines of this region may have witnessed some of these outbreaks; it seems certain, however, that nothing of the kind has transpired since the discovery of America. The purported finding of corn, pieces of pottery and old ruins imbedded in lava has been investigated by the writer on several occasions and in every instance it was a clear case of "mistaken identity."

Many of the rivers of New Mexico have been deflected from their former courses by being obstructed and entirely dammed by the more recent lava flows. The most noted example of this nature is found in the Rio Grande, extending from the upper part of the Española valley north sixty miles to the San Luis valley in Colorado.

Throughout this entire distance the river channel was completely filled by the vast lava sheet which extended over an area of at least ten thousand square miles. It is difficult to conjecture the position of the old channel; it may have been as far west as Ojo Caliente.

Throughout the whole sixty miles exists a grand gorge, carved through a level plateau, with almost perpendicular walls rising majestically over one thousand feet.

To this obstruction in the Rio Grande the origin of the fertile San Luis valley is due. That this valley was an old lake bed there is not a shadow of doubt; the final opening of the gorge drained this vast Pleistocene lake, and the valley of San Luis was thus born.

The Española valley had its origin in a similar manner, due to damming at the railroad bridge; not far above Albuquerque obstructions of this nature at one time existed.

Fig. 2.—THE "ENCHANTED" MESA. Photographed by W. M. Borrowdale, 1900.

All of the beautiful valleys along the Gila and San Francisco rivers were formed in like manner; the old lake bottom exists in a number of distinct terraces, which were formed at different periods, as the outlet of the river gorge was lowered. It is not improbable that the former course of the San Francisco river may have been down Duck Creek to the Gila at Cliff; the nature of Duck Creek Valley and the difference in levels of the two rivers seem to favor this supposition.

The two principal classes of formations most generally found on the surface over the territory, are the lavas (*mal pais*) and Cretaceous rocks.

In New Mexico it is thought that about one-sixth of the land surface is covered with eruptive sheets and fully one-third of the whole area by Cretaceous formations.

The minerals of economic importance most generally sought in New Mexico, are associated with the older classes of eruptive rocks and later gravels; in the massive Carboniferous limestone; and in the later or uppermost Cretaceous sandstones.

This association of particular classes of minerals with special types of rock formation, is now recognized as a matter of fact by every intelligent person.

In being able to recognize mineral bearing horizons and formations, is a qualification much to be desired in the successful pursuit of economic geology.

A generalized vertical geological section of the earth's crust in New Mexico is given for comparison with similar sections in other states and elsewhere.

It is seen that the geological column is fairly complete; the principal exception being the lower Paleozoic rock system. At Lake Valley, Tierra Blanca and near Silver City, it is thought that portions of the Silurian and Devonian exist; confirmation on this point, however, is lacking. Any omissions of this character that may occur are partly offset by a very full development of the Permo-Carboniferous series. The sequences of the "red beds" series are very complete, and probably reach a total thickness of 3,500 feet in certain localities, which includes the Jura-Triassic systems. The "red series" are conspicuous for their wide distribution of copper, disseminated through certain strata of shale and sandstone. Since the

Loam and Sand
River Gravel, Talus wash
Lava
Santa Fe Marls
Loose Sands.

Puerco Series
Gray Shales & Clays
Sandstone
Coal
Fire Clay
Fox Hills Coal
Sandstone
Shales, &c.

Drab Gypsiferous
Shales
Clay & Slates
Sandy Shales and
Limestone
Dark Shales
Buff Sandstones
Limestone
Shales
Vermillon Beds
Red Clays & Loose S.S.
Chocolate Beds
Sandstones
Gypsum Beds
Oklahoman Series
Red & Variegated
Shales
and Sandstones
Compact
Gray
and
Blue Limestone
Mag. Limestone
Slates & Shales
Lacking?
do.
do.
Quartzites

Schists

Gneiss   Granite

Recent		Qt.	Cenozoic
Pleistocene			
Neocene		Ter.	
Eocene			
Laramie		Cretaceous	Mesozoic
Montana			
Colorado			
Dakota			
Jurassic		J.-T.	
Triassic			
Permian or Permo-Carb.		Carboniferous	Paleozoic
Carboniferous			
Devonian? Silurian? Ordovician?			
Cambrian			
Algonkian			
Archean			

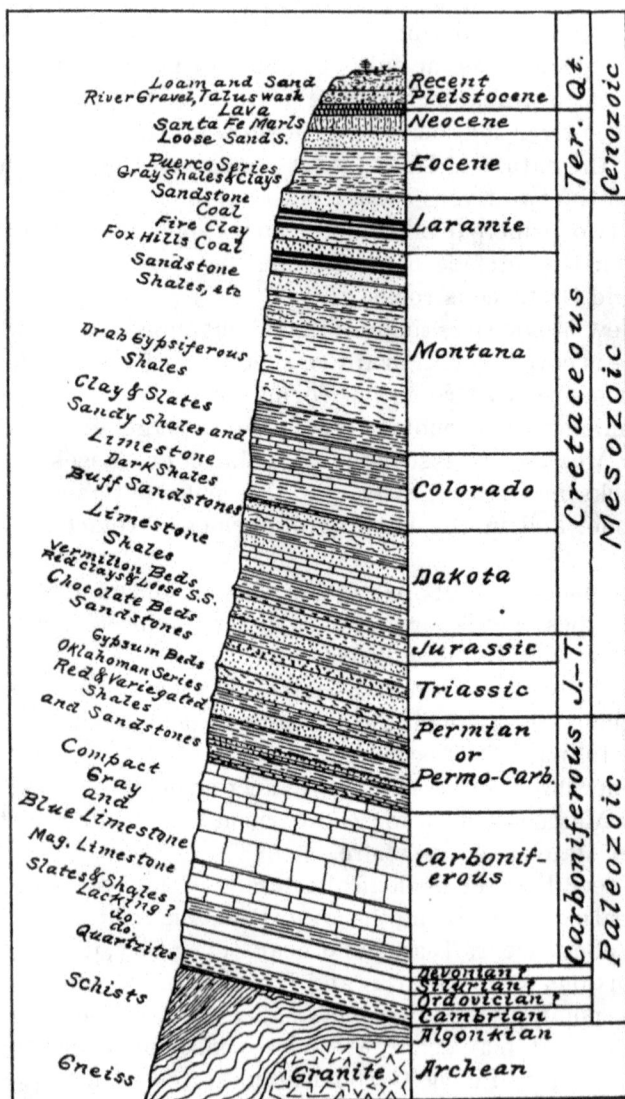

Fig. 3—VERTICAL SECTION OF THE EARTH'S CRUST
IN NEW MEXICO.   By F. A. Jones.

copper is very low grade, but few localities will pay to work. Copper in many instances has replaced fossil plants and trees in the "red series" deposits; this character of ore is a high grade glance.

The coal deposits of New Mexico all lie in the upper Cretaceous sandstones; some beds are in the Laramie and others belong to the Fox Hills series, or upper Montana.

Only two principal fields are definitely known to belong to the Fox Hills series; these are the Cerrillos and Carthage fields.

It is observed that the Fox Hills coals are superior in quality to the later Laramie deposits; since their coking qualities are much more pronounced. Whether this desirable feature is due to the superiority in age, or whether the conditions were more favorable from being influenced by eruptive members, or both, is a matter that will require further investigation.

The Carboniferous limestones are recognized as mineral carriers, or ore bearing horizons. Much of the copper and all of the principal lead and zinc ores are found in intimate association with this important series.

Gold and silver ores are most generally or always found intimately associated with the metamorphic and eruptive types of rocks; a good gold section is usually recognized by the intelligent prospector as a region having an abundance of "porphyry." To a greater or lesser degree the associated minerals of lead, copper, zinc, etc., are found intermixed with the gold and silver ores.

The occurrence of the rarer metals, is usually in association with the more plentiful commercial metals; and are frequently extracted as a by-product.

In the stream and river sands of the Territory intimately connected with the placer gold and black sand, platinum and a number of the heavy and rare metals unquestionably exist in a greater or lesser degree. Very little attention was ever given any of the concentrated black sand, and our knowledge of its contents is mainly conjectural.

CHAPTER II.

EARLY SPANISH CONQUESTS FOR GOLD.

Shortly after the discovery of America, legends of the glory and riches of the new world agitated the whole of Europe. The successful conquests of Mexico and Peru for gold under the leaderships of Cortes and Pizarro followed in quick succession. Spanish desire for wealth and adventure at this period was irrepressible. New fields were sought; attention finally was directed toward the mysterious land of the north —New Mexico.

In the United States of America, New Mexico is the cradle of the first conquests for the precious metals.

Of the many early expeditions of adventure, conquest and discovery, Alvar Nuñez Cabeza de Vaca and his three companions were the first Europeans to set foot on New Mexico soil, A. D. 1534. They were, indeed, the true discoverers: having approached the Territory from the east. This discovery may be regarded as one of chance, and not of conquest; since Cabeza de Vaca was trying to reach European civilization through channels by way of the west, in order to escape servitude from the coastal tribes of the Gulf of Mexico, where he had been held in captivity for several years, having become stranded on that coast in a disastrous shipwreck. Cabeza de Vaca speaks of turquoise which he saw at his farthest point north in the Rio Grande valley, presumably near the present site of Bernalillo.

In 1539, Friar Marcos de Niza, on hearing of Cabeza de Vaca's sojourn to the strange country in the north, at once fitted out an expedition and went as far as Cibola (Zuñi). The conquest of this friar was more of a religious nature than for the aggrandizement of wealth. He speaks of seeing turquoise and some gold during this expedition.

The most noted of these early expeditions was made by the Spanish general Francisco Vasquez de Coronado and his army; this was an army of conquest for gold.

Coronado took pretty much the same route as Friar Marcos de Niza and arrived at Cibola (Zuñi) on the 9th day of July, A. D. 1540.

There can be no doubt about Cibola being practically the old Zuñi Pueblo, of western McKinley county, near the Arizona line.

A short time after the arrival at Cibola, Coronado despatched Alvarado, a captain in his army, to explore the country south and east. The exact route taken by Captain Alvarado is somewhat obscure; but he drifted to the southeast, touching the Rio Grande near La Joya then following up that stream to a point at or near the present site of Bernalillo. Here Alvarado was met by his general and all went into winter quarters, until the following spring, among the Pueblos along the Rio Grande.

In May, 1541, Coronado began his memorable march to the east and north in search of the famed city of Quivira (not Gran Quivira), reaching that city about the middle of the summer. The only metal seen at Quivira was a small piece of copper which the chief of the tribe wore about his neck; this metal no doubt came from some of the ancient workings of mines in Michigan.

After spending some days at Quivira, feeling keenly the disappointment in not finding treasures in gold and silver, he returned with his followers to the Rio Grande, the place where they had passed the winter.

A second winter was spent in this rendezvous: during the meantime short expeditions, under the command of captains, were sent out to explore the contiguous territory.

Some turquoise and gold were realized though not to the full extent of their expectations.

Not meeting with the success in securing gold, silver, and precious gems that they anticipated before starting on the expedition, disappointed and disgusted at his failure, Coronado called his forces together and quit the country in 1542, by way of the same route that he had entered the Territory. Thus ended the first great hunt and conquest for gold in New Mexico.

From a humanitarian point of view, with respect to the treatment of the aborigines by the invaders, Coronado's army might be considered nothing more than organized highway robbery, rapine and murder perpetrated on a peaceful and harmless people.

Some forty years after the expedition of Coronado, Don Antonio de Espejo visited Cibola and other places and came across several Christian Indians at Cibola, who had accompanied Coronado into this strange land from Mexico and had remained there ever since. Those Christianized Indians related stories to Espejo's men of populous cities on the banks of a great lake, far to the west of Cibola, where gold was so plentiful that the women wore it in the form of bracelets and charms.

Espejo returned to the Rio Grande, after his visit to Cibola and the silver mines in Arizona, and made a journey to the northeast from some point on the river. This brief record says: "Here they were informed of the rich mines of the precious metals, some of which they visited and took from them good glittering ore."

Oñate, like Coronado, excited over the legends of fabulous wealth lying to the far northeast, started in June 1601, for Quivira. As he proceeded on his journey, strange stories of great cities of this golden empire came to his ears from time to time, which greatly stimulated the hopes of his army of adventurers. The inhabitants were said to have utensils of the precious metals and decorated their persons with ornaments of burnished gold.

After chasing this phantom of imagination for some months, Oñate returned to San Juan about the first part of October, chagrined and disappointed like the irrepressible commander who preceded him thither, by some sixty years.

In 1620, Padre Geronimo de Zarate Salmerom speaks with delight of New Mexico's climate, agriculture and mines, in doing missionary work at that time among the Pueblos.

Apparently disgusted and feeling much remorse over the cruelty of his countrymen toward the aboriginal tribes, this missionary declared that the Spaniard would enter the "doors of hell" in order to satisfy his craving for gold, were it possible to obtain gold from that region.

Many expeditions were planned and executed under various leaders from about 1580 to the Pueblo uprising a century later;

several of these conquests were of considerable historic interest, whilst others were mere incidents of adventure.

Narratives of fabulous wealth of some mystic Eldorado, were to the conquering Spaniard, always a stimulus for further adventure and conquest.

Stories of lost or hidden treasurers always have a charm of romance and mystery woven about them, and it is never difficult to find believers in such traditions.

It is said that a treasure of $3,000,000 in gold is buried beneath the ruins of the old church of the Gran Quivira in eastern Socorro county. This vast fortune, it is claimed, was in the custody of the hierarchy at the time the city was destroyed by a volcanic eruption. This ancient Pueblo was likely contemporaneous with the principal cities of the Rio Grande valley, and passed into decadence from the same causes that obliterated most every other aboriginal Pueblo in the west. From the nature of the ruins, especially that of the old church, it is not at all improbable that a great portion of this city was of Spanish as well as Pueblo origin.

The extinct volcanic crater, situate at the north end of the great lava flow (locally known as the *mal pais*), is what has been suggested as the source which dealt destruction to the city. This crater is some thirty-five miles to the south and a little east of the Gran Quivira. The writer has examined into this matter critically and can see absolutely no evidence that would in any way connect the destruction of the Pueblo with the volcano. It is possible that the city may have been abandoned, due to the effect of seismic disturbances, which caused the water to sink in its natural basin, from whence the inhabitants formerly derived their supply; since one of the old basins of Pueblo times, some six miles to the southwest, is used by a ranchman at the present time; the impervious adobe floor with the imprint of fingers was plainly to be seen.

The evidence is such as would indicate that the volcanic flow which formed the *mal pais* must have taken place at least three centuries before the advent of the Spaniards, and in all probability at a still earlier date.

The ruins of the Gran Quivira have been dug into in many places, at various times, by different persons, who have been lured to that desolate spot by mythical traditions, in search

Fig. 4—RUINS OF THE GRAN QUIVIRA. Photographed by F. A. Jones, 1903.

for the buried treasure. Not a dollar in gold or silver has ever come from those old ruins, so far as the writer has been able to find out.

Some mining in a desultory manner, has no doubt been carried on in New Mexico, under Spanish* rule, and a little, perhaps, done by the aborigines. Yet the evidence, aside from the turquoise mines at Los Cerrillos and the Burro mountains is sufficient to satisfy the statement that no true metal mining was ever carried on within the borders of New Mexico until about the beginning of the year 1800; with the possible exception of *Mina del Tierra* in the vicinity of the turquoise near Los Cerrillos.

Governor Chacon in 1803, says: "Copper is abundant, and apparently rich, but no mines are worked."

Lieutenant Pike in 1807, speaks of only one mine in New Mexico; the inference being, no doubt, the copper mine at Santa Rita.

Pino in 1812 mentioned that old silver mines had been found closed up with the tools inside; these workings were perhaps nothing more than prospects and had been done by the Spaniards prior to the Pueblo revolt in 1680. The evidence

---

*The uprising of the Pueblo Indians in 1680 was said to have been due to the hardships of slavery inflicted on those people by the Spaniards, principally in working the mines. Be this as it may, history tells us that the Spaniards had to flee the country as a result of their extreme cruelty and barbarity. About the close of the 17th century they returned, and it was expressly stipulated by the Pueblo Indians that they should not again engage in mining, but only in agricultural pursuits.

It seems that the Jesuits were, formerly, the principal miners before the revolt of 1680, and they were the ones on whom the Pueblos mainly wreaked their vengeance in a general massacre.

Tradition has it that all the known mines at that time were filled in and so completely covered up, that when the Spaniards returned and who were prohibited from working mining, the succeeding generations were unable to discover the mines again. During the temporary exodus of the Spaniard from the country, the records were either destroyed by the Pueblos or were carried to Mexico and Spain by the fugitives in their flight, resulting in the complete obliteration of all the early mines and mining records of the country.

From all the evidence obtainable on the early Spanish mines, we are forced to the conclusion that the mining in that day was scarcely removed from what we now term prospecting.

The writer has had exceptional opportunities to examine into all the purported Spanish mines in the Territory, and must confess that, with the exceptions of the turquoise mines and *Mina del Tierra*, nothing else has been seen that would approach the resemblance of a mine up to the present time.

In the archives at Santa Fe, under date of 1713, is the document of *Nuestra Sra. de los Reyes de Linares*, which refers to an old covered up mine in *Sierra de San Lazora*. (Old Placer Mountain). And in 1714 a gift of the whole or part of this same mine was made.

After this date about the year 1800, the *Santa Rita* mine was discovered and worked. Then followed the real discovery of the Old Placers in 1828. On December, 1833, the *Santa Rosalia* grant was recorded in favor of *Jose Francisco Ortiz*, in the *Sierra de Oro*, west of *Mina del Compromiso*, (north and south). The following day December 19th, 1833, a record of a mine was made adjoining the Ortiz mine. The New Placers were discovered in 1839.

In 1846, the *Santo Niño*, north of the Ortiz mine, (Old Placer), is recorded: as also, in the same year, a mine in the *Real San Francisco del Tuerto*.

This practically constitutes most everything pertaining to the records of the earlier mines.

seems conclusive that no mines of either silver or gold were
worked to any extent prior to 1800; save some little gold picked
from the gravels at various points throughout the Territory
and from the silver lead mines in the vicinity of Los Cerrillos,
so alluded to above.

Under Spanish rule prospecting for placer gold was carried
on to a certain extent; yet no rich finds were ever brought to
notice, excepting at the Old and New Placers.

There would be no good reason to claim that the Pueblo
Indians or the early Spanish explorers were better qualified
to find rich mines than the modern prospector of today.

The Spaniard has been a gold hunter from the earliest times
and placer gold was the kind he knew most about; lode mines
were not so alluring to him.

There are a number of old workings in New Mexico, of
limited extent and presumably of Spanish origin, which have
been discovered by the modern prospector; but the richness
of the ore or deposits has been almost invariably disappointing.

It might be added here that the traditional stories of lost
mines are the *ignes fatui* that have held many a prospector
spell-bound and carried him into unknown regions, ultimately
resulting in giving to the world a Cripple Creek or a Klondike.

The enchanted Adam's diggings, the legendary Peg-leg
lode, the mythical Log Cabin mine and similar stories of lost
lodes, exist in imagination only; yet, they serve as a stimulus
to the prospector who with pick and pan paves the way for
civilization. Such fantasies when viewed from an unprejudiced
standpoint, are to be regarded as real and necessary factors
in the successful hunt for gold.

CHAPTER III.

PLACERS.

It should be noted in the foregoing chapter that the placer fields of New Mexico were the first to come into notice in the mining of the precious metals and were exploited to a certain extent by the Spaniards, and also, perhaps, in a few instances, some gold was taken from the superficial gravel beds in the vicinity of the the Ortiz mountains and some of the streams by the Pueblo Indians, prior to the coming of the Spaniards. In the latter instance this would not properly come under mining as such finds would be classed as purely accidental.

Before describing any of the gold bearing lodes, a general idea of the extent and nature of the placer fields of the Territory should first be briefly considered.

The magnitude of the placer area of New Mexico is but little comprehended by the unobserving person. The area will approximate four hundred and fifty square miles; the richness of which will average not less than sixteen cents per cubic yard of gravel throughout the entire deposit.

It is true that the greater portion of these vast fields may never be successfully worked; yet we dare not circumscribe the workable area, since modern engineering skill and genius would defy a fixed limit. It is no exaggeration in saying that most every handful of sand and gravel from any of the streams and gulches of New Mexico carries gold values.

The writer, when Director of the School of Mines of New Mexico, made a number of delicate experiments along this line, of the stream sands taken from near the surface deposits of the Rio Grande and the Galisteo river and only in a few instances was the absence of gold noted.

The enormous erosion which has been in course of operation throughout the past geological ages, due to the uplifting of the great western plateau region, was the means of liberating the particles of gold and concentrating the same in the beds of the various gulches and streams. Owing to the great specific gravity of gold it naturally accumulates at the lowest points, in pockets and riffles close to or on bed rock; and it is there that we should expect to find the richest deposits.

From the foregoing it is observed that the Rio Grande basin is evidently the largest receptacle for such accumulations, and undoubtedly contains more gold than any other stream in New Mexico.

The enormous wealth lying in the bed of this stream cannot be arrived at, even approximately.

In speaking of the placers of the Rio Grande, above Española, Professor Silliman says : "Here are countless millions of tons of rich gold quartz reduced by the great forces of nature to a condition ready for the hydraulic process, while the entire bed of the Rio Grande for forty miles is a sluice on the bars of which the gold derived from the wearing away of the gravel banks has been accumulating for countless ages, and now lies ready for extraction by the most improved methods of river mining. The thickness of the Rio Grande gravels often exceeds six hundred feet, or three times that of like beds in California, while the average value per cubic yard is believed to be greater than in other accumulations yet discovered."

The thickness of the gravel beds at places not contiguous to the Rio Grande is often very considerable. An artesian well was sunk at Taos a few years ago to a depth of over four hundred feet, which was in gravel the whole distance. The drillings from this well were panned at intervals during the progress of the work, gold being found in every pan.

The Moreno valley at Elizabethtown at one time was a lake, several hundred feet deep; but, since the Cimarron river has cut through the dike, forming the gateway to the valley, the lake was thus drained, leaving a deposit of sediment from twenty-five to three hundred feet thick. Every place where bed rock or hard pan has been reached in this valley plain, gold has been found in considerable quantities. The large dredge of the Oro Dredging Company, operating in the Moreno river about one mile below Elizabethtown, is meeting with remarkably good success. This river flows through the north half of the Moreno valley and forms a junction near the gateway with the Cieneguilla river which drains the south end of the valley. The Moreno river, being rich in gold has for ages deposited all its sediment in the Moreno lake basin; there can be no question or doubt, whatever, about the richness of the auriferous sands and gravels of the old lake bed.

The richness and extent of the Moreno placers, embracing those of Willow creek and Ute creek, the latter lying on the east and opposite side of Baldy mountain, are not paralleled by any other deposits in the Territory. Further mention will be made of the placers about Baldy mountain, when the lode mines are discussed, in that section of the country.

The extensive area of auriferous sands and gravels which surround the basal slope of the Ortiz mountains have been almost continuously worked since their discovery in 1828. The new placers at Golden may also be included with the deposits of the Ortiz mountains, which were discovered eleven years later.

Describing a quadrant, with the point where the Galisteo is intercepted by the Rio Grande as its center, and with a radius of twenty-five miles, most all of the area embraced in this quadrantal sector, may be considered placer ground.

The whole of the channel of the Galisteo river is one vast sluice box; the gold collecting in the depressions and seams of the rocks that form the river bed. The same thing will apply to the Rio Grande; the only difference is that in the latter sluicing is being conducted on a more gigantic scale.

A section to bed rock of the Rio Grande showing the different strata passed through, with the assay value thereof, per cubic yard, is shown in Fig. 5. This section is taken from Professor Arthur Lakes' hand book on "Prospecting for Gold and Silver in North America."

This probing of the river bed was done by the Santa Fe Placer Company under the direction of Professor E. Walters, geologist; on account of the strong sub-flow the shaft was sunk by the aid of the pneumatic caisson. The value per cubic yard at bed rock was found to be $6.25.

The trough of the Rio Grande is thus seen to be a vast storehouse of gold.

The Oro Grande Company of Pennsylvania, with J. P. Rinker of Tres Piedras, manager, has begun the construction of a dredge to work the auriferous sands and gravels in the bed of the Rio Grande, above Cieneguilla, in Taos county.

The river bed for several miles at this point is considered favorable for dredging; since bed rock or hard pan in most places is only from two to ten feet deep. The river is very narrow and all the water is confined to a restricted channel, which will be especially beneficial in floating the dredge.

Owing to large boulders of *mal pais* which are said to cover the river bottom at every few feet, this company may meet with some very serious trouble in coping with the same; with

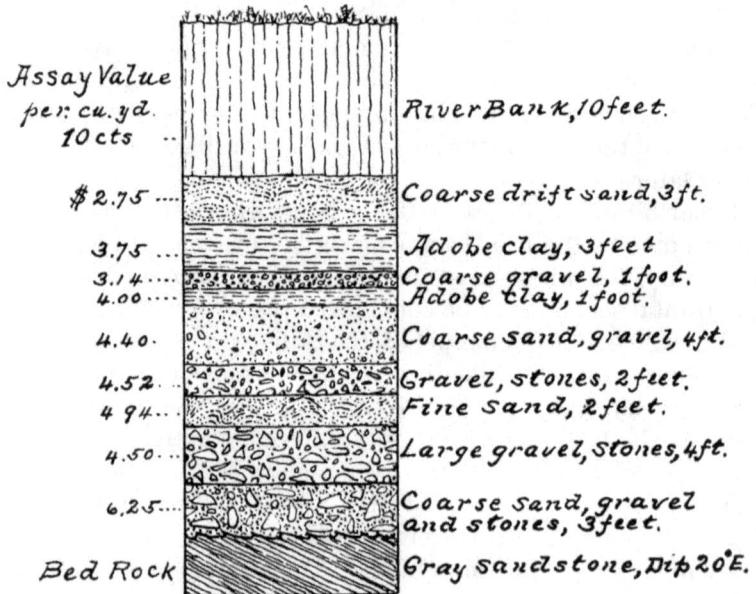

Assay Value
per. cu. yd.
10 cts

River Bank, 10 feet.

$2.75 .... Coarse drift sand, 3 ft.

3.75 ... Adobe clay, 3 feet
3.14 .... Coarse gravel, 1 foot.
4.00 .... Adobe clay, 1 foot.

4.40 . Coarse sand, gravel, 4 ft.

4.52 . Gravel, stones, 2 feet.
4 94 .. Fine sand, 2 feet.

4.50 .. Large gravel, stones, 4 ft.

6.25 .... Coarse sand, gravel and stones, 3 feet.

Bed Rock   Gray sandstone, Dip 20° E.

Fig. 5—VERTICAL SECTION OF THE RIO GRANDE BASIN.

this difficulty overcome, the other operations are to be regarded as favorable to dredging.

These *mal pais* boulders frequently attain enormous proportions; they having fallen into the river gorge by the gradual recession of the cliffs, which are very precipitous on either side of the river. This grand gorge of the Rio Grande extends from a point about five miles below Embudo north to the San Luis Valley, a distance of about sixty miles.

The river has considerable fall all through that distance and bed rock is very shallow.

Much or all of the placer gold along the river here evidently had its origin in the Sangre de Cristo range of mountains. Tributary to this locality are the Rio Hondo placers, said to have been worked by the early Spanish explorers, under the Jesuits.

The placers around Pinos Altos are inaccessible from any large supply of water, and must be worked on a smaller scale than those of the Moreno valley or Rio Grande. The same is also true of the Jicarilla gravels in Lincoln county, north of White Oaks.

The newly discovered placer fields of the Pittsburg mining district, lying at the base of the Caballo mountains, in Apache canyon, next to the Rio Grande, have the advantage of being on or near the river; but, the water must be elevated several hundred feet to reach the rich ground should mining be done on an extensive scale. This, however, is a problem to be solved by mining engineers, should exploitation of those fields prove them to be sufficiently rich in gold, to justify the installation of a large pumping plant. At Jarilla, in Otero county, good ground exists, although the field is quite limited in area.

The excitement over the find of the Caballo mountain placers had scarcely subsided when a dispatch reported that a new gold field was discovered on January 20, 1904, eighteen miles east of Tucumcari, on Revuelto creek, Quay county. A few days later the find was declared a hoax and that the ground had been salted.

The Las Animas placers at Slap-jack hill and in the depression and flats below have produced more gold by dry processes than most any other dry camp in New Mexico.

This field lies north of Hillsboro and is producing at the present time about $450 per month; the gold being taken out by the primitive methods of the Mexican.

At one time, shortly after the discovery of gold in this section, hydraulic mining was carried on to a certain degree of success; but owing to the inadequate water supply, which was piped from considerable distance, this method was abandoned.

Near Good Hope, in Rio Arriba county, an hydraulic plant is being installed by J. P. Gill, and it is expected to begin work

at a very early day.   These placer grounds are rather circumscribed in their superficial extent.

A few miles above Abiquiu, on the Rio Chama, there exists
an extensive area of river sands and gravels which have been
prospected by some two or three companies with favorable
results.

In 1900 some Kansas City and St. Louis capitalists tried to
work the ground, but failed to make a success of the enterprise. The failure was due to the process adopted, the dredges
used, being constructed on wrong mechanical principles and
were not able to handle the material, or even save the values
after the gravels were excavated.

This ground will average about twenty-three cents per cubic
yard on the benches on either side of the river and as much
as $2.20 per yard on the bed of the stream.

These gravel beds range from twenty-five to forty feet in
thickness, and have their origin, in part, in Colorado, on the
east side of the continental divide.   Much black sand exists in
this gravel which carries good values in gold and perhaps,
some platinum.

George S. Du Bois has recently discovered placer gold near
the head of the Las Animas creek in Sierra county, the
discovery was made during March, 1904.

In other portions of the Territory aside from those mentioned, placer mining is being conducted successfully in a
small way by Mexicans; and still in other places where the
prospects are quite favorable, the conditions for lack of water,
wood and transportation, bar anything whatever from being
done.

CHAPTER IV.

NEW PLACER (Silver Butte) DISTRICT.

This district lies to the south of Cerrillos, a town on the Atchison, Topeka & Santa Fe railway, in Santa Fe county, and near the west line thereof.

Most mining districts in New Mexico are very indefinite in regard to their extent or area; generally embracing a whole cluster or range of mountains or continuous mineralized belts, regardless of size or shape.

Thus it is with the New Placers; they are supposed to include everything to the plains each way, from the north slope of the Ortiz mountains south, to the plains south of South mountain. This embraces the Ortiz mountains, the Old Placers, Dolores, Golden, New Placers, San Pedro and South mountain. As a gold mining district this is the oldest in New Mexico; it is also noted for its recent production of copper.

Nuggets of gold were no doubt picked up occasionally in this area, by the Pueblo Indians; though no real mining was ever conducted in this field by those people, so far as any evidence can be obtained. It was in the year 1828 when gold was first discovered in this district. The point of discovery is what is known as the "Old Placers;" and was made by a herder from Sonora. It is said that some of his herd strayed into the Ortiz mountains whither he went in search of it; seeing a stone which he thought resembled some of the gold bearing rocks of Sonora, he examined it and the rock proved to be rich in gold.

News of the discovery soon spread and the excitement was intense. The most crude appliances imaginable were used; notwithstanding, considerable gold was taken out. Winter seemed to be the most favored time for mining; by melting the snow with hot rocks they were able to work until the dry season of the year. The gold was washed or panned out in a "batea," a sort of round wooden bowl, about the same diameter of the modern gold pan. The mode of operation was first to fill the "batea" with the auriferous sands and gravels, and then immersing the whole in water and by constant stirring and agitation, the mass of sands and gravels was reduced until nothing but black sands and particles of gold remained in the wooden vessel; this mass of black sands and gold was then reduced in a clay retort to obtain existing values, after the largest nuggets and particles of gold were first removed.

According to Prince's History of New Mexico, between $60,000 and $80,000 in gold was taken out annually between the years 1832 and 1835. The poorest years about this period were from $30,000 to $40,000.

About this time an order was given prohibiting any person from working the mines excepting the natives. Foreign capital and energy was thus excluded, which greatly hampered and handicapped development. Under this new regime, each Mexican miner held one claim, the size of which was ten paces in all directions from the main discovery pit. Any claim not kept alive by labor after a certain length of time, was subject to relocation.

The gold was mainly in nuggets and dust; one nugget* claimed to have been found was worth $3,400, which netted the finder only $1,400. If true, this was the largest nugget ever discovered in New Mexico. The fineness of this gold is about 918. It would be hard to estimate the exact amount of gold taken from the "Old Placers," but it must have been considerable.

Thomas A. Edison, the celebrated American inventor, erected in 1900 a large plant at Dolores to operate on these rich gravels; after making some experimental runs the plant was closed down indefinitely. The process was held a secret.

Much rich ground yet exists in this section; but owing to the Ortiz grant having passed into the hands of a syndicate, which holds it under a 99-year lease, little or no work has been done of late. Apparently, the policy of the lessee is obstructive to the interests of the district, as it will not allow prospecting of any kind on the property; neither will it sub-lease.

This land grant covers all of the Ortiz mountain and the best part of the placer grounds of the district; it embraces an area

*The size of the nugget thus reported is much to be doubted.

of ten square miles, having the Old Ortiz mine as the center of the grant.

In 1833 a vein of gold bearing quartz was discovered on the Ortiz property, which was on the famous *Sierra del Oro*, and now known as the Ortiz mine.* The owner of the property took into partnership a Spaniard by the name of Lopez, a person well skilled in mining of that day.

Through the management of Lopez, their mining operations were successful and a considerable sum of money was realized.

Wishing to retain the full production of the mine, through jealousy and cupidity for gold, Ortiz sought a channel to rid himself of his Spanish colleague. The plan was carried out

Fig. 6 - ORTIZ MINE; the Shaded Portion Shows Extent of Mexican Stopes Prior to 1850. From an Old Map; Scale 200 Feet to the Inch.

under the pretense of an obsolete decree which forbade any Castillian from residing or operating in New Mexico.

Accordingly, Lopez was forced to leave the country. Ortiz then formed a copartnership with several of the rascally officials, who were connected with the expulsion of Lopez, and proceeded to work the mine.

The new management not being familiar with mining operations were wholly unsuccessful; history tells us that they did not obtain "one grain of gold."

This famous historical mine has been worked at intervals ever since its discovery, recent years excepted. The vein,

---

*The claim made by some that the Ortiz mine is the oldest lode mine in America is a mistake; *Mina del Tierra*, in the Cerrillos district, exceeds it by 100 years, at least. In fact the Santa Rita mine is a quarter of a century older. Don Cano, the discoverer of the Ortiz mine, came to Mexico from Spain in the early part of the 19th century.

apparently, is enclosed in syenite-porphyry; its strike is N. 13° E. and its dip is 75° toward the northwest.

The vein outcropping is an oxidized iron-stained quartz; below the depth of 85 feet the ore becomes base, carrying sulphurets of both iron and copper.

The top portion of the vein was first worked out on account of its free milling qualities.

The New Mexico Mining Company which acquired the Ortiz grant in 1864, was first organized in 1853 and incorporated in 1858.

In 1865 this company began the erection of a 20-stamp mill which was completed in the early part of the year following. This stamp mill was the first erected in New Mexico.

A certain degree of success crowned the efforts of this company; and in 1869 it began adding an additional 20-stamps to its plant.

The ore was conveyed from the mine to the mill by means of a tramway. After a few intermittent mill runs the mine closed down; bad management the cause.

Some years later another company erected a large amalgamating and concentrating plant at the mine, which was never operated successfully.

The Cunningham mine, in Cunningham gulch, near Dolores, is also well and favorably known to the district. This is among the earliest locations of the district; belonging now to the Sandia Gold Mining and Milling Company.

The outcropping is immense; the width of the vein is about 600 feet and can be traced for a long distance. So bold is the outcropping that it can scarcely be classed as a vein; but more properly what miners term a "blowout."

The whole of this mineralized dike consists of quartz and feldspar with rich seams or streaks passing through it in various directions. The quartz is more or less stained with oxide of iron at and near the surface; with depth the ore becomes refractory. The hanging wall is a syenite-porphyry and the foot wall a quartz-porphyry. The dip and strike of this lode conform with the Ortiz vein.

Among other and familiar lodes may be mentioned the Candelaria, belonging to the Galisteo Company; the Brehm lode originally worked under the management of the New

Mexico Mining Company which owned the Ortiz mine, the Hutchason lode,* the Brown lode, and the Humboldt 100th; the latter lode named in honor of Humboldt's centennials. The Shoshone is also a prominent lode which has been more recently located. All of the above lodes lie near Dolores and the gold from the Old Placers evidently came from these veins, due to the action of erosion.

The New Placers from which the district takes its name is situate some four or five miles to the south of the Old Placers in the Tuerto (San Pedro) mountains. This new field was discovered in 1839; eleven years later than the Old Placers.

Much gold has been taken from the gulches at this place; the San Lazarus gulch is quite a steady producer at the present time.

In the vicinity of Golden, which is the newest part of the placer district, much activity is manifested, and considerable success attends the efforts of modern mining. The gravels in this section average from twenty-five cents to one dollar per yard of material handled. Scarcity of water, as at the Old Placers, is a serious obstacle in working this ground. The fineness of the gold is about 920.

Concerning the geology of the New Placer district it seems that the trio—South mountain, Tuertos (San Pedro) and Ortiz mountains—are most intimately connected in their origin and had their birth in one common disturbance. The orographic line of weakness was north and south; on this line the three pustules of syenite-porphyry broke through the horizontal sedimentary capping of the overlying Carboniferous and Cretaceous series. Generally speaking the topography of these groups is identical.

South mountain is not so familiar to the general public as the other two groups; inasmuch, that section appears to be less mineralized than the Tuerto and Ortiz localities.

At the Tuertos (San Pedro), which are about three miles north of South mountain, the sedimentary series have been partly elevated and dip about 15° toward the east. The Oroquai mountain which is the eastern member of the Tuertos is entirely stripped of any former sedimentary covering,

---

*Col. J. S. Hutchason (Old Hutch), the discoverer of the Magdalena district, located this old time mine. Col. Hutchason was in the Old Placer district as early as 1848, and at one time owned the Candelaria mine.

exposing the rugged character of the syenite-porphyry; having its counterpart in the Ortiz peaks, some four miles to the north.

The now deserted village of Dolores stands to the northeast from the Ortiz mountains, near its base.

Gold, silver, lead, copper iron and zinc are found in this district.

In the classification of the mode of ocurrence of the ores, three divisions would seem proper :

(1) Deposition due to erosion, placer gravels.

(2) Deposition due to descending and ascending waters and the filling in of fractured zones and true fissures, which carry gold.

(3) Deposition due to contact metamorphism, from which the copper, lead, silver and zinc ores are intimately associated.

In the first of these divisions the placer gold has its origin in the universally accepted manner ascribed to such deposits: that is, through disintegration of the rock-complex of the second classification, as above given.

Since there appears to be two distinct features which characterize the occurrence of the gold under the second division, the veins are divided into fractured zones and true fissures. The first of these has no banded structure and the walls are undefined, greatly crushed and shattered. In the second case a true banded appearance is recognized while the walls are definite and intact. It would appear from a close inspection of the two classes of veins that the first was filled by a leaching process of descending waters; some of the seams and pockets have proven immensely rich in gold. But, in following this shattered zone down, the values grow less as the crevices grow smaller. Sulphides usually appear from seventy-five to one hundred feet below the surface. Eventually, the fracture becomes so small at increased depth, as to disappear altogether and the vein is completely lost or said to have "pinched out." These crushed and shattered mineralized zones are by far the most numerous of any types of deposit in the district. The relative position of their planes approaches perpendicularity, their general strike is nearly east and west.

Under the true fissure class of veins only one or two of any consequence have been noted. The most prominent of this class is found in the famous Ortiz mine. This vein is completely encased in syenite-porphyry and has a banded appearance. Ascending waters or lateral secretion is responsible for the mineralization of this and similar lodes of the district. Some very fine specimens of leaf and wire gold have been taken from the various properties. Beautiful specimens enclosed in calcite have been found in the Gold Standard mine.

Deposition under the third and last division is the most important in the district when viewed from a commercial standpoint. Here may plainly be seen the effect of pneumatolytic action, induced by the porphyritic magma, which was forced upward against the Carboniferous limestones. The writer was much impressed by the effects wrought in the overlying sedimentaries, by this intrusive eruptive, during a recent visit at the mines of the Santa Fe Gold and Copper Company. This property is, by far, the best developed of any in the district; the workings are quite extensive, embracing several miles of development. The ore is principally of a low grade chalcopyrite, and intimately associated with garnet, lime and shales.

Massive limestones in some places have been converted into garnet, exceeding one hundred feet in thickness, in some instances. The superficial limestones and shales at the copper mines are frequently penetrated by andesite dikes. It was observed that the best ore bodies were found at or just above the main porphyrite contact and along the contact planes of the andesite dikes. From the foregoing, it would appear that the segregation of ores, along or near these planes of contact, is largely if not wholly due to the action of aqueous, acid and gaseous vapors in their effort to escape from their magmatic prison; under released pressure their metallic burden was thus, necessarily dropped.

At the Lincoln Lucky mine the deposition of ore, no doubt, was similarly induced by the porphyrite intrusive beneath. Since the ore occurs in limestone along a shattered zone and not in direct contact with the porphyry, this view, at first, does not seem well taken. Upon closer investigation it will be found that cavities in the limestone have been mineralized, only where communication with the igneous member existed.

On the eastern and northeastern slopes of the Tuertos are some iron properties which have not yet been fully exploited; the Perry group being the most prominent.

Some of the principal lode claims are: The San Lazarus, Gold Standard, McKinley, Lincoln Lucky, Anaconda group, Stockton group, Alto group, San Miguel, Gold King group, Hazelton group, Shamrock group, Lucas mines, New Mexico group, Good Enough groups (San Lazarus gulch), and the Old Reliable (on the Ortiz grant.)

The more prominent of the placer properties may be enumerated as the Monte Cristo Mining Company, Santa Secivel, Baird Mining Company, Ltd., Morning Glory and Gold Dust, the Red Bank and Viola.

CHAPTER V.

CERRILLOS (Galisteo) DISTRICT.

Under the caption of Turquoise, something in a general way as well as special, is said concerning this famous mining district.

From a historical standpoint, no section in the United States is possessed of so much interest. The profound ancient workings at Mount Chalchihuitl, due to the existence of turquoise in that locality, seem almost incredible that the work could have been accomplished with the crude appliances of the stone age, and yet, such was the case.

Fragments of coiled pottery, stone hammers, lichen covered rocks and trees over a century old growing on the old dumps and in the working pits, when first brought to the notice of American explorers over fifty years ago were then hoary with age and prove beyond the shadow of doubt the great antiquity of mining in this region.

This celebrated district lies on the north side of the Atchison, Topeka & Santa Fe Railway, at the little village of Los Cerrillos, near the center of Santa Fe county.

The first description of the region was given by Prof. W. P. Blake, who visited the old turquoise workings in 1858. Prof. Blake's article was published during the same year in the American Journal of Science. Other distinguished scientists and writers paid visits to that section, prior to the modern discovery of metallic ores.

It was in the year 1879 when the modern prospector drifted into the region after the great excitement at Leadville, Colorado. The discovery of sulphide ores, zinc, lead and silver, was heralded abroad and the boom started.

Two town sites, Bonanza City and Carbonateville, were staked out in the early '80s. and a tidal wave of mining craze swept over the district.

These once thriving villages are now scarcely more than piles of rubbish and fallen walls. It was in the old hotel at

Carbonateville, some of the walls are yet standing, where General Lew Wallace, when seeking recreation in the mining camp also read some of the proof sheets of Ben Hur.

Beside the ancient turquoise mines, there exists a metal mine which was worked for its silver and lead, and is almost as old as the *Chalchihuitl* workings; this is known as *Mina del Tierra*. In this mine exists the only real evidence of ancient lode mining in the southwest; it antedates the first work done in the Ortiz and Santa Rita mines by at least a century.

The old working consists of an incline shaft of 150 feet and connects with a somewhat vertical shaft of about 100 feet in depth. Extensive drifts of 300 feet connect with various chambers or stopes; these chambers were formed by stoping or mining out the richer ore bodies.

The full extent of this old working has never been definitely determined; since the lower depths are covered with water which would have to be pumped out to fully explore the mine. As late as 1870 the remains of an old canoe were still in evidence, which was used for crossing water in the mine or, as a carrier for conveying the waste and ore to the main shaft; from this latter point it was carried to the surface on the backs of Indians in raw hide buckets, or "tanates."

The shaft had step-platforms or landings every twelve or fourteen feet which were gained by climbing a notched pole (chicken ladder), similar to what some of the Pueblo Indians use at the present day. Many crude and curious relics, such as stone hammers and sledges, fragments of pottery, etc., have been taken from both the mine and the dump.

It is thought that the Jesuits had this work performed by Indian slaves prior to 1680.

The labor involved, when we take into consideration the crude manner of doing the work, is something tremendous. Throughout this district are a number of (smaller pits and openings which are thought to have been done at that time, from the association of similar crude implements found about the works.

The ore from this mine is a sulphide of lead and zinc, carrying rather high values in silver. Silver was, no doubt, the principal metal sought and utilized.

A smelting plant of two stacks, one for lead and the other

for copper, of 50-tons each, was erected in 1902, at Los Cerrillos, on the railroad, but was never operated steadily. The ores of the district, without first making a separation of the lead from the zinc, cannot be successfully smelted at a profit. The Cash Entry, Grand Central and Tom Paine mines have been more extensively developed than most of the other properties, and are credited with some production.

The Golden Eagle, M. & L., J. B. Weaver, Galena Chief, Fairview, Sucker Boy, Evelyn group, Astor group, Empire State, Beta, Little Joe, Sunnyside, Whalen group and Ingersoll constitute the principal claims.

There were fully one thousand locations made during the primary impulse of the excitement.

The ores of the district are heavy sulphides of zinc and lead, carrying some silver and a little copper and gold. The region is thoroughly mineralized and on the west is traversed by numerous andesite and basalt dikes.

The central core of the district about Grand Central mountain, is an augite-andesite porphyry; and in the region of the turquoise mines, at both Chalchihuitl and Turquesa, it is much altered by kaolinization.

Immediately east of the augite-andesite area, embracing the arroyo of San Marcos, the porphyry is recognized as a hornblende-andesite. Since the andesite formation embraces all of the metal mines in the district, it is attributed as being the chief carrier of the metaliferous values.

This mineralized area is traversed by innumerable veins and veinlets more or less irregular but all having a general strike of about N. 30° E.

It would seem that the cause of the numerous systems of veins and veinlets that abound in the district, was due to the cooling of the andesitic magma, which resulted in extensive checking and fracturing in adjusting itself to the changed condition.

Escaping gases and aqueous vapors in their effort to escape along the lines of least resistance, deposited their metallic burden under released pressure. In addition to this phenomena, circulating waters at a later period must have also given aid in the segregation of the metallic sulphides along these fractured zones.

Fig. 7—LOS CERRILLOS SMELTER. Photographed by F. A. Jones, 1902.

A valuable contribution to the scientific literature on the Cerrillos district is "The Geology of the Cerrillos Hills," by Prof. D. W. Johnson, formerly of the University of New Mexico, which appeared as a reprint from the Columbia School of Mines Quarterly, during 1903.

CHAPTER VI.

CENTRAL DISTRICT.

This district lies immediately southeast of Pinos Altos and seven miles east of Silver City. Due north two miles is the military reservation of Fort Bayard.

Central district embraces the sub-districts of Hanover, Fierro, Santa Rita and other outlying points.

Practically contemporaneous with the discovery of gold at Pinos Altos this district sprang into existence; the copper mines at Santa Rita, however, were known as early as 1800. Originally, the village of Central was called Santa Clara. As early as September 20, 1869, there were recorded in the Central district at Fort Bayard, fifty-seven mining claims. Most all of these locations were made at Hanover, Santa Rita and San Jose; only a few were made in the vicinity of Fort Bayard.

In the locality of Central post office, and more especially farther to the northwest, the country is somewhat broken and its surface is more or less covered with a light colored sand rock. Porphyry dikes have shouldered their way upward breaking through the thin covering of the sedimentary series. The exposed sedimentaries are of Carboniferous age. The veins are contacts and fissures, and their mineralization is due to contact metamorphism and pneumatolytic action. Since the first work in the early sixties (barring Santa Rita, Hanover and Fierro), that portion of the district about Central post office has remained quiet, with the exception of an occasional ripple of excitement caused by some find of minor importance.

It was on the 16th day of June, 1903, when the camp was suddenly revived by a phenomenally rich strike in Gold Gulch on the Pactolus claim.

The discovery was made by two miners, Saunders and Cornell, practically on the line between the two fractional claims of the Pactolus and Owl. These were old claims, having been located a number of years, and the strike was a very great surprise to the owners.

As usual, excitement ran high, and the adjacent country was relocated for quite a distance in all directions from the point of the discovery. Several new companies have been formed and development will likely uncover other ore bodies.

The ore is of a very refractory nature carrying considerable zinc; in appearance it is quite deceptive.

By far the most noted mine in the Central district, and perhaps in the United States, historically considered, is the renowned copper mine of Santa Rita.

This mine was discovered by an Indian in the latter part of the eighteenth century, who afterward revealed his secret in 1800, to Lieutenant Colonel Manuel Carrasco. Colonel Carrasco was a commandant in the Spanish army, who had charge of the military posts at that time throughout certain portions of New Mexico. It seems that Carrasco made no attempt to work the property during the time it was in his possession. A wealthy merchant, Don Francisco Manuel Elguea, of Chihuahua, purchased the property in 1804 from Carrasco, and immediately set a force of laborers to work developing the property.* Absolute title was not really vested in the Spanish Colonel; for the Indians gave him permission to work the mine only under certain conditions.

Lieutenant Pike in 1807, during his expedition to the Territory, in speaking of mining says: "There are no mines known in the province, except one of copper, situated in a mountain on the west side of the Rio del Norte, in latitude 34°. It is worked and produces 20,000 mule loads of copper annually. It contains gold, but not quite sufficient to pay for its extraction."

There can be little doubt that the mine referred to by Pike was the Santa Rita mine, since Don Francisco, the merchant of Chihuahua, was operating the property at that time; although, the latitude of 34° would be too far north to correspond with Santa Rita.

---

*According to Prince's history, Carrasco found the copper of such fine quality that he contracted the whole production to the royal mint for coinage.

The metal was transported principally on the backs of pack mules from the mine to the City of Mexico.

One hundred mules carried three hundred pounds each and were said to have been kept constantly employed.

Fig. 8—VIEW OF SANTA RITA MINING CAMP.

James O. Pattie, a trapper at that time, in his narrative concerning the property says: "The mine was worked by a Spanish superintendent, Juan Onis, for the Spanish owner, Francisco Pablo Lagera.

Within the circumference of three miles there is a mine of copper, gold and silver, besides a cliff of lodestone.* The silver mine is not worked, not being so profitable as either copper or gold mines. The Indians were very troublesome and the trappers did good service in keeping them in order, by force and treaties."

It appears that Pattie's Francisco Pablo Lagera must be the same person that Prince's history gives as Don Francisco Manuel Elguea, the Chihuahua merchant.

In R. W. Raymond's Report on the Statistics of Mines and Mining for the year 1870, page 403, gives the name of Don Francisco de Alquea as being the person who purchased the mine from Colonel Carrasco. Dr. Raymond gives the name of the mine superintendent at that time as Ori, which corresponds to the time of the man Onis in Pattie's narrative.

In 1809, the death of Don Francisco occurred; the property was then worked under lease by Juan Onis or Ori, through a contract by Don Francisco's widow.

After working the property several years Juan Onis was superseded by the two Pattie brothers who leased the property a number of years, paying $1,000 per annum. The elder of the two brothers remained at the mines and established a ranch on the Mimbres a few miles away and prospered. Finally in 1827 the elder Pattie concluded to purchase the property agreeing to pay $30,000 in gold therefor. The money was entrusted to a pretended Spanish agent who claimed to represent the owner, and instead of turning over the money as was expected, nothing was ever heard of this impostor again. The rascally trick of the pretended agent virtually ruined Pattie financially.

It seems that the widow of Don Francisco disposed of the mines about the year 1822; the purchaser, in 1826, was exiled as a Spaniard and the property passed into the hands of Robert McKnight, who held it from 1826 to 1834. It is said

---

*Evidently the iron deposits at Fierro, now being worked by the Colorado Fuel & Iron Company.

that McKnight realized a considerable profit during the eight years he had possession.

On account of hostilities of the Apaches, the mine was abandoned for a few years. About 1840, Siqueiros took possession and worked the property continuously until 1860. During the year 1860, Sweet and La Coste purchased the mine and worked it up to the time of the Confederate invasion in 1862, when all of the mines in the Territory closed.

During the interval between 1862 and 1870, Sweet and La Coste interested two other persons having the names of Brand and Fresh. This quartette under the name of Sweet, La Coste, Brand and Fresh, employed a number of Mexicans from Chihuahua, paying them $2.00 per day in Mexican coin. They operated a small Mexican blast furnace having a capacity of about 2,000 pounds of copper per month. This metallic copper was freighted by wagon to Sheridan, the terminus of the Kansas Pacific railway; from that point it was sent to the eastern markets.

The present ownership of this historical mine is the Santa Rita Mining Company, which company purchased the property from J. Parker Whitney in May, 1899. Just how long Whitney held the property and from whom he purchased, the writer was unable to ascertain.

The estimated production of the Santa Rita property from the time of its discovery to January 1, 1904, is approximately 80,000,000 pounds of metallic copper.

The Santa Rita mines occupy a depression or basin, resembling a vast ancient crater. At the south side of the basin, Whitewater (Copper mine) creek cuts through the rim and affords a drain to the depression.

The rock in which the principal copper values lie is a feldspathic porphyry, which has reached a rather advanced stage of kaolinization; the state of decomposition of this porphyry is more marked in some places than in others.

It seems that the prophyry on cooling, or perhaps from extraenous forces, has become checked and cracked in every conceivable direction. It is in these openings in which the bulk of the copper value is deposited. On account of the enormous quantities of metallic (native) copper disseminated in and through this felsite porphyry, we are led to believe that

secondary enrichment has taken place. Chalcocite is dis·
cernable in the felsite rocks and it is from this mineral and
also perhaps, from chalcopyrite, that the native copper chiefly
owes its origin. Surface waters appear to have been the
sole carrier of the metalliferous values, which were arrested
during descending infiltration. Most all of the associated
ores of copper are found in the camp; cuprite, however, seems
to be quite plentiful, incrusting the native copper.

The principal part of the Santa Rita camp is owned by the
Santa Rita Mining Company; the greater part of the mining
is done by a system of leasing.

Many interesting traditions are interwoven in the en·
chanted name "Santa Rita."

To the east of the Santa Rita basin ou the rim is a peculiar
isolated column of stone which rises to a considerable height,
and may be seen from certain directions for long distances.
By a little imagination the stone resembles a woman kneeling,
in the attitude of prayer; this monolith is known as the
"kneeling nun." Superstition has clothed this silent sentry
of stone in the following abbreviated legend :

"In the early days of the Spanish conquest of Mexico, upon
the mountain there stood a mission or cloister, wherein dwelt
monks and nuns; and one of the latter, a Sister Rita, a nun
professed, who had broken her vows, was turned to the stone
or monolith now standing on its brow."

The old Spanish prison in which the peons or slaves were
confined for disobedience in working the mines, is still
standing.

Two of the adobe forts known as the "Martello" towers, are
still standing and in a fairly good state of preservation.
These towers are circular; the inside diameter is twelve feet
with an equal height, and the walls are three feet thick.
There were four of these originally; one standing at each
corner of the prison yard, with port holes near the top. These
towers proved impregnable against the assaults of the Apache
Indians, when the mines were being operated in the early
days.

Passing across the low divide from the Santa Rita, in
going west, the Hanover gulch is encountered. The postoffices

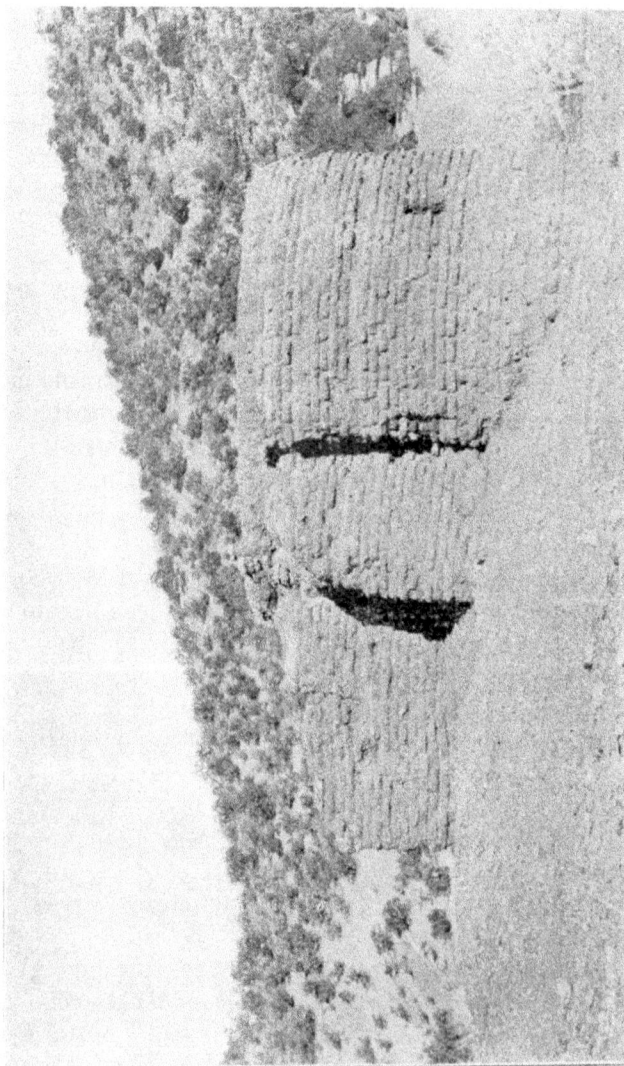

Fig 9—OLD SPANISH ADOBE PRISON.

of Hanover and Fierro lie in the gulch; the latter lying about two miles north of the former.

Near Fierro is the celebrated Hanover mine; exceeded in production and renown, only by that of Santa Rita. This mine was, perhaps, known to the Spaniards about 1804; but its true discovery was made by a German, from Hanover, in the early or middle fifties. According to Dr. R. W. Raymond, in the Statistics of Mines and Mining of 1870, this property had a greater production when in operation, than the Santa Rita mine and had been more or less a producer since 1858. Between 1858 and 1861 it is credited with a production of 1,000,000 pounds of metallic copper. The cost of mining and smelting this copper was ten cents per pound. The smelting of the ore was done in a high blast furnace, and the metal was refined in a reverberatory furnace.

Before the war of the rebellion, the copper was run into pigs weighing from 100 pounds to 120 pounds and hauled by mule teams, via Mesilla to the Texan port of Lavacca at a cost of six cents per pound. From thence it went by schooners to the City of New York, for $5.00 per ton.

At the time of the Confederate invasion, work on the mine suddenly stopped; all the machinery and equipments, including 187,000 pounds of copper, were taken to San Antonio, Texas and confiscated by the southern troops. After the period of invasion, the mine passed into a state of "innocuous desuetude," for a number of years.

T. B. Catron, of Santa Fe, finally became the possessor, and from him it passed to C. F. Grayson and company, of Silver City. This latter company sold the property to Phelps, Dodge and Company, in 1902, the owner at the present time.

Following the Santa Rita range to the southwest about three miles from the Santa Rita mines the old San Jose copper mines are reached.

Malachite and azurite are the chief ores of this property, and are associated with a quartz gangue. The strike of the veins here is about N. 20 E. cutting a porphyritic formation. Many years ago these mines were said to have been profitably worked; they were abandoned prior to 1870, but during 1902 some production was reported.

Central district with its numerous sub-districts and camps

Fig. 10—OLD ADOBE FURNACE AT HANOVER. Erected in the Early Fifties of the Past Century.

is by far the most important mining section in New Mexico. Chiefly to this district is due the credit of placing Grant county at the head of the mineral producing counties of the Territory.

Among the important minerals and ores found in this section are copper, iron, zinc, lead, gold and silver. No coal has been found in Grant county.

It is 'not within the sphere of this little volume to even attempt a description or discussion of the mines in this district. The fact, however, should not be overlooked that nearly three fourths of the metallic wealth of New Mexico, for 1902, was from Grant county—the greater part of which came from the Central mining district.

Prominent among the various properties at Fierro are the Anson S. and Iron Head, the two latter are controlled by the Colorado Fuel & Iron Company; the Copper Queen, Modoc and Hanover (the latter mine is described above), owned by the Phelps-Dodge people; the Emma, Hanover No. 2, Nora, Dude, Holy Moses group, and a great many others.

The Santa Rita Mining Company controls the principal producing claims of the Santa Rita basin. The Log Cabin and Belmont were producers from that section in 1902; these claims are chiefly lead properties. Down Whitewater creek about a mile is the Wild Cat property, which is familiar to the district.

Lying east of the post office of Hanover, about half a mile is the celebrated zinc mine known as the Thunderbolt. Much sulphide and carbonate ore has been shipped from this property of late years to Mineral Point, Wisconsin, for treatment. In the vicinity of this post office are the Copper Queen, Minnie B., Philadelphia, Copper Kettle, Copper George, Peacock and the Hanover Iron mine.

At the post office of Central, besides the claims already mentioned in the new district at Gold gulch, a number of companies are doing development work. The St. Louis Gold Gulch Company was organized in September, 1903, and own the Lucky Bill and Dutch Uncle claims.

The most noted property in the vicinity of Central post office is the Texas. This property has produced considerable in the past; the ore is a sulphide carrying heavy values in silver with a little gold. The vein is a contact between slate, shale and porphyry. The Missouri, Jasper and Helene are producing properties and are classed as producers in 1902.

Lone Mountain District.

This district is properly a sub-district of Central, it lies south and west of Central post office between four and five miles. There are four patented properties; it is a silver camp. Very little, if anything, is doing in the camp at this date.

Lone Mountain district was discovered in 1871 by Frank Bisbee (from whose name Bisbee, Arizona, was christened) and Jack Frost. A 10-stamp mill was immediately erected and operated on the rich silver ores there for about two years, before closing down indefinitely.

Mimbres District.

This district lies a few miles to the northeast of Santa Rita in Grant county, and embraces the once lively camp of Georgetown, extending beyond into the Mimbres valley.

The discovery of silver here dates back to the year 1866; the pioneers were Messrs. Butine and Streeter, George Duncan and Andy Johnson and a few others. Afterward came D. C. Casey, Jas. Fresh, E. Meeks, Lige Wicks, S. S. Brannon, David Smith, Alex. McGregor and Charles Nicolai.

It was several years later before any work was attempted; the first development was on the McNulty location, done by Jas. Fresh and E. Meeks.

In 1873 it began to dawn upon those who had locations that the camp was one of considerable merit. A few years later witnessed a veritable boom and the greatest excitement prevailed. The ore from which most of the values came was cerargyrite (horn silver) found in a greenish slate or shale contact with the overlying carboniferous limestones.

The plane of contact approached horizontality and the deposits were practically of blanket form.

Occasionally sulphides and chlorides of silver are found, sparingly intermingled with galena; notably among this class of mines is the Jackson group. Vanadinite is found in considerable quantities in several of the mines, as well as some ruby silver.

The present owner of the Quien Sabe is G. B. Sibole; W. H. Bentz owns the Silver Bell group. The Commercial mine was very prominent as a producer in the early days.

In another part of the district is the Naiad Queen group, a patented property, embracing 199.47 acres, and belonging to the Mimbres Consolidated Mining Company, and was the most prominent property in the district. Many other properties with a record of production, exist in the district, though now idle.

On entering Georgetown late in the afternoon on April 23, 1903, the writer and his companion were much depressed by the awful stillness that pervaded the premises. In fact, absolutely nothing was found doing, the streets were depopulated and grown up in weeds. Long rows of buildings casting their ghostly shadows by the lingering sun, impressed us with a feeling of indescribable fear and horror. The once bustling moving throng of sturdy prospectors and miners who had "struck it rich," the incessant clattering of the stamps in the silver mills and the sharp crack of the mule driver's whip, all have been forever silenced in the brief span of a decade, by the magic touch of time. Oh, what utter desolation! The flitting picture before us is a realistic view through the kinematoscope of the past—it is the passing of a western mining camp. At the end of these series of depressing views, we behold towering above the wreckage and piles of waste a beautiful monument of solid silver, glinting in the setting sun, representing a production of $3,500,000 to the credit of the camp.

Carpenter District.

Near the south end and on the west slope of the rugged Mimbres range of mountains, in Grant county, is an isolated mining district of which but little is known to the outside world. This section is covered with a dense pine forest and the topography is so broken that prospectors usually find it convenient to shun, in their still hunt for fortune.

The deposits are principally contacts between limestone and porphyry, carrying the sulphides and carbonates of zinc and lead.

It is surmised by the writer that the district is destined to become one of the great zinc fields of New Mexico, when properly exploited.

The Grand Central group appears to have merit in its deposits of smithsonite, as well as the Potosi group, which carries sphalerite along with smithsonite. The Beanie lode is a lead silver proposition and contains but little zinc.

CHAPTER VII.

PINOS ALTOS DISTRICT.

The early history of this district is very similar in all respects to that of others in the west.

Not only had the pioneer to overcome vast distances and endure intense suffering, due to thirst, heat, hunger and cold, in pushing his journey into an unknown region; but, had to be constantly on the alert, since he was hunted in the day and haunted at night by the savage hordes which infested the land.

Next to the discovery of gold at Dolores and Tuerto, of the Old and New Placers in Santa Fe county, came the Pinos Altos finds, now in Grant county.

Gold was discovered* at Pinos Altos in May, 1860, by a party of '49ers, who drifted into the country from California. Three persons composed the party, Col. Snively, Birch and Hicks; Snively was the recognized leader. The discovery was made by Birch, while taking a drink out of Bear gulch just above its junction with Little Cherry, near the Mountain Key mill. This place was called Birchville, in honor of the discoverer. It is claimed by some that the first find was in Rich gulch, near the present site of Pinos Altos; be this as it may it was Snively's party that made the discovery.

By June, quite a number of prospectors had gathered to the new Eldorado and were busily engaged in washing out gold. The gold fever became so contagious that within an incredibly short time it had infected a number of people in Texas, California, Sonora, Chihuahua and Missouri; in December fully 1,500 persons were at the diggings.

From $10 to $15 per day, per man, was realized in the gulches in the immediate vicinity of the original finds.

Mr. Thomas Mastin in December, 1860, discovered and

*It is claimed that there are records among the archives of the Mexican government in the City of Chihuahua which speak of gold being found in the Pinos Altos mountains in the beginning of the 19th century. The discovery is said to have been made by Gen. Pedro Almendares, one of the commandants of the Mexican outpost at Santa Rita. If true, no practical results came from it.

located the first quartz lode. This lode is the present Pacific mine and crosses the "continental divide," where the waters come to the "parting of the ways." This noted property was bought by Virgil Mastin, a brother, in the following spring.

A few lode claims were located in 1861; among them was the Lock vein, since made famous as a gold producer under the name of Mountain Key.

During the winter and spring of 1861, the Apache Indians constantly menaced the life and property of the miners. In the fall, on September 27th, a severe engagement took place between the miners and a band of 500 Indians, under the famous Apache leaders, Mangas Coloradas and Cochise. The miners were ultimately victorious, but Captain Thomas Mastin, who commanded a company of volunteers lost his life, and several others, during the bloody conflict.

After this engagement most of the people through fear, quit the country, only a few of the more reckless remained behind; Virgil Mastin being one of the number who refused to leave, in order to avenge the death of his brother, should an opportunity be presented. Several years later Virgil Mastin was ambushed and killed near the Silver Cell mine.

But little work was done during 1861-4*, as most all the Americans had abandoned the camp.

During this interval of abandonment, the Mexicans changed the name from Birchville to Pinos Altos. Owing to the forest of "tall pines" which existed there at that time the name Pinos Altos was very suggestive and the latter appellation has clung to the place ever since.

About the close of 1864 the camp was attaining its former prestige by an influx of American miners and mining was again on the eve of prosperity when another raid was made by the Apaches, who succeeded in terrifying the inhabitants and driving off all their cattle and horses.

Nothing further was attempted in mining until 1866, when the Pinos Altos Mining Company was organized and chartered under the laws of New Mexico. The members of the organization were Virgil Mastin, J. Edgar Griggs, S. J. Jones, Joseph Reynolds and J. Amberg.

---

*The governor in his report of 1861-2, alludes to the fact that 30 gold lodes at Pinos Altos were working, employing 300 men, and the ore was worth from $10 to $250 per ton.

By July, 1867, this company had completed a 15-stamp mill which was the second quartz mill erected in the Territory; this mill was preceded only about two months by the one at the old Ortiz mine, in Santa Fe county.

Other mills followed at short intervals afterward and during the years 1868-9, great activity was witnessed throughout the whole district.

The records show that by September 19th, 1869, there were located and recorded 213 quartz lodes, beside those of the placer claims.

In 1872, Skillicorn and company built and put in operation the well known Mud Turtle mill, so called from the fact that a large mudturtle had been captured at that place.

During the year 1883 Peter Wagner built a 5-stamp mill and Place and Johnson also put up a 10-stamp, during the same year. This latter mill was not a success on account of having to deal with refractory ores.

The mill that Wagner built was provided with a concentrator, and to him belongs the first honors of being able to successfully handle refractory ores in New Mexico. To Wagner is due the credit of building the first concentrator for the concentration of base ores in the Territory.

In 1887 Lunger and company discovered high grade ore in the Mountain Key mine; which was shortly afterward purchased by General Boyle. Boyle organized a company and erected a 20-stamp mill and shortly afterward took out $500,000. This property laid idle about a decade until the owner W. C. Chandler, in April, 1903, began operations and the property has been producing ever since.

Among the celebrated properties of the Pinos Altos district, the famous Silver Cell mine deserves especial mention.

This property lies southeast of Pinos Altos about two miles, and such high grade silver in a gold camp is looked upon as an anomaly.

The discovery was made on June 18, 1891, by the three Dimmick brothers, who had emigrated from Pennsylvania and taken up a homestead in the vicinity of their future find.

One of the brothers on driving up the cows which were grazing in the adjacent hills, incidentally picked up a stone, as he supposed, and tossed it at one of the animals that had

Fig. 11 · SHAMROCK SMELTER.

fallen out of line. Just as the missile passed from his hand his attention was attracted by its heft.

Following the supposed stone in the direction it had gone and on picking it up, to his great astonishment, it was found to be solid silver. Reporting and exhibiting his find to his two brothers, whereupon a search was immediately inaugurated by the trio, resulting in the discovery of the lode.

The discovery transformed the business of the three brothers from that of dairying to mining. Systematic development was prosecuted and up to January 1, 1903, the property had produced $100,000.

In March, 1903, the property passed into the hands of the Shamrock Gold and Silver Company. The present company are finding free silver ore at a depth of 400 feet.

A smelting plant has been installed on the property, which treats not only the ores from the Silver Cell, but from the surrounding camps of Pinos Altos, Central, Hanover and Santa Rita.

The Silver Cell lode seems to be that of a green stone or diorite dike, intruded into the granites. The minerals are native silver, argentite and cerargyrite; very little, if any gold is present.

On the west side of the mountains from Pinos Altos about two and one-half miles is located the Cleveland group of five claims.

This is among the oldest exploited properties in the district. It is claimed that J. Amberg, a German metallurgist, did considerable work on this property, prior to the time that he became interested in the organization of the Pinos Altos Mining Company, in 1866.

The formations seem to be a series of syenite and quartz-porphyries, associated with some intrusives of diorite, with andesite coverings.

The general trend of the ore bodies is to the northeast and the dip averages about 30° toward the northwest. Heavy sulphides of copper, iron, zinc and lead, carrying values in both gold and silver, is the character of the ore. Zinc-blende always shows more prominently at increased depth throughout the district.

Several properties which are associated with the history of

the district have extensive development, and it is not within the scope of this volume to attempt separate descriptions of each.

The basal core of the Pinos Altos district, is reddish and grey colored granite and gneiss of Cambrian age, penetrated by diorite-porphyry dikes; the veins, are thus fissures in the main.

The placer gold evidently resulted from the disintegration of the quartz seams and porphyritic dikes which traverse the district, and could not have been transported any appreciable distance.

Moreover, the form of the gold and the altitude at which it is found in the mountain gulches corroborate this statement.

The fineness of the placer gold is 775. Most of the placer mining is done in a small way by the Mexicans. Panning, dry washing and the use of the arrastra are the methods employed. Of the placer claims the Log Cabin and Adobe are the most important.

Among the most prominent lode claims, beside those hereinbefore mentioned are the Pacific Extension, Atlantic, Deep Down, Aztec, Manhattan, Mammoth, Gopher, Arizona, Nogal, Blue Horse, Gold Bell, Alaska, Tom Boy, Dover, Ribbon, Nugget, St. Louis, Comstock, Maud S., San Pablo, Hardscrabble, Golcondia, Hilltop group, Little Pacific, Golden Giant, Esperanza and Pride-of-the-West.

The production of the Pinos Altos district from the time of its discovery to January 1, 1904, will approximate $4,700,000.

CHAPTER VIII.

SILVER OR CHLORIDE FLAT (Silver City) DISTRICT.

Adjoining the town of Silver City on the west is claimed to be the place where silver was first mined* in New Mexico. In this particular locality very little silver was taken out, when compared to what was produced at Silver or Chloride Flat. Owing to the phenomenal finds of silver here, which point is about one and one-half miles from town, Silver City received its name.

Approximately, $3,250,000 in silver were taken from this circumscribed area in a comparatively short time.

The discoverers of Silver or Chloride Flat were Jim and John Bullard†, J. R. Johnson, John Swishelm and others, in the spring of 1871.

A number of mills, the Wisconsin company, M. W. Bremen, Tennessee Mining and Milling Company, Cibola Milling Company and the Carrasco smelter, were in operation at Silver City during the active days of the district.

Between six and seven miles further to the northwest in the same mining region is Camp Fleming, named in honor of J. W. Fleming, of Silver City, who is one of the pioneer miners, and was one of the principal operators in that section at that time. The discovery of Silver in Camp Fleming was due to the finding of a piece of very rich float by Dutch Henry and partner, in 1882. The production of this camp is not known, but was considerable.

The ores of the district occur in the same limestone belt as Chloride Flat, along the contact. Argentite and cerargyrite were the two principal classes of silver ores.

---

*The first modern silver mining was done here; the ancient mining at *Mina del Tierra* at Los Cerrillos is to be excepted in this statement.

†John Bullard was killed the following winter by an Apache Indian who was mortally wounded in the back, and who leveled a pistol in both hands, lying on his belly, shot Bullard through the heart.
In this skirmish Bullard had emptied his gun and pistols, when he beheld the Indian in the act of firing; calling to a companion, "shoot that Indian quick." His companion's gun snapped and did not fire and the next instant Bullard fell forward on his face, dead.
Immediately afterward the Indian rolled over and expired. Bullards peak, west of Silver City, was named after the hero.

The region about Camp Fleming is sometimes spoken of as Bear mountain; it is now deserted. The Granite Hill group is at Bear mountain.

White Signal (Cow Spring) District.

This section lies a few miles southeast of the Burro mountain district and seems to possess some merit in becoming a producer of turquoise, as well as gold, silver, lead and copper.

None of the properties are very extensively developed; notwithstanding the excellent surface showings.

The formation is much the same in nature as that in the Burro mountain district.

The principal properties are the Coplen group of four claims, J. W. Carter mine, and the locations owned, five claims, by the Michigan-New Mexico Copper Company.

The district of White Signal received its name from a large white porphyry rock which is conspicuous in that region.

Bullard's Peak (Black Hawk) District.

This district is situated in the north end of the Burro mountains and is the north extension of the Burro district.

It received its name from John Bullard who was killed in that region in 1871, by the Apache Indians.

A negro by the name of Bowman, or better known to the old timers as "Cherokee Jim," was the first discoverer of rich silver float in the district; this was in 1881.

On the strength of "Cherokee Jim's" find, John Black and a partner by the name of Sloan, discovered and located the Blue Bell (now Alhambra) mine, the source of the rich float found by the negro.

Shortly after the location of the Blue Bell (Alhambra) the Rose and Black Hawk were discovered and located in the order named.

A man by the name of Shedd soon afterward bought up practically every location made at that time.

The Alhambra (formerly Blue Bell) lode seems to be a contact fissure occurring between syenite and a peculiar porphyry; the width of the vein is remarkably uniform, varying from two to four feet, and well defined. A talcose vein-stuff containing some quartz gangue, fills the fissure; the whole of this product is fairly well mineralized.

Native silver and argentite constitute the character of the ore, which occurs in narrow rich shoots and frequently runs as high as 15,000 ounces per ton. A depth of about 400 feet is attained on the Alhambra with drifts to the same extent.

The Solid Silver Mining Company has eight patented claims; the principal location of the group, and the one on which the most development has been done, is the Black Hawk. The records of the company show that the mine has produced nearly $600,000 in silver; one car load of the ore is said to have brought almost $28,000.

About 750 feet of work constitute the development; the character of the formation is much the same as at the Alhambra.

Next in importance is the Hobson group; the vein is nearly four feet wide on an average and a depth of about 300 feet has been reached.

Clark's Peak District.

Clark's peak district is somewhat obscure; it lies west of Silver City about 30 miles, near a prominent peak called, Clark, after the discoverer of mineral in that section. The region is more definitely located by referring it to township 18, south, and range 18, west, which is to the northeast of the Anderson district.

There are four patented properties here and the ores carry gold, silver, copper and iron.

Nothing, absolutely, is doing in the district.

Burro Mountain District.

On account of the Apache Indians the Burro Mountain district never received more than passing notice until in the later '70s. and early '80s.

The district lies about fifteen miles southwest of Silver City, in Grant county.

It seems that two brothers, Robert and John Metcalfe, were the first to make locations in the region, in about the year 1871.

At that time these mountains were the strong hold of the

savage Apaches and it was a common belief that no prospector
ever returned who chanced to wander into the region. The
beautiful Mangas valley to the northeast received its name
from the celebrated Apache chieftain, Mangas Coloradas,
whose deeds are written in blood on the pages of frontier
history.

The "old timer" John E. Coleman, better known in the
early days as "Turquoise John," is generally credited as being
the discoverer of the district, since nothing of any consequence was ever done before his time. He was in the
district in 1879 and made a number of locations of both copper
and turquoise, having discovered turquoise in some ancient
workings during that year. The years 1882-5 witnessed
considerable activity throughout the district and the better
part of the ground was taken up at that time. During this
active period two small smelters were erected, the Paschal
smelter and one other; these were the first installations of
modern appliances in the district. Neither of the plants
proved successful and the camp was condemned for a while
on this account. Everything lay dormant until 1900, when
the district was revived by more careful and systematic
prospecting and mining.

A 100-ton concentrating plant is being erected at the St.
Louis mine and the recently erected Ben Johnson smelter in
"Dead Man's" canyon has been torn down and removed to
Silver City on account of the lack of water and fluxes.

Johnson's smelter is now in successful operation at Silver
City and a second smelter is being erected on the site of the
old Silver City reduction works, by the Comanche Mining and
Smelting Company.

Among some of the more familiar properties of the district
are: The Klondike, St. Louis group, Virginia, King and
Queen, Comanche group, Jo E. Sheridan group, Carter
group, Morrill group, Favorite, Samson group, Silver City
mine, Hazel group, Fannie, Tarantula group, Santa Ana
group, Connecticut mine and the Amazon group; beside
the many valuable mines of turquoise.

A majority of the copper properties are now consolidated
under the management of the Burro Mountain Copper Company and one or two other organizations of like character.

In the Burro mountain district proper, the chief metallic ore is copper; on it and the turquoise (see chapter on turquoise) depend the success of the region. The principal copper-bearing portion of the Burro mountains, covers an area of three miles in length by about two miles in breadth.

Only in a brief and general way has anything ever been written or done concerning the geology of the district.

The general strike of the formation is observed to be in an easterly and westerly direction. On the east the formations are largely covered by andesite breccias and conglomerates, which have apparently come from the eruptive region about Bullard's peak, which lies a few miles to the north.

Granite and trachyte porphyries are the characteristic rock features; these are traversed by numerous small quartzite dikes. Metasomatic action has produced extensive kaolinization of the feldspars throughout the various portions of the district. It is chiefly in these koalinized rocks that the most valuable turquoise of the region is found. The metallic ore bodies consist of a low grade copper carbonate, in which the seams and seamlets of the rocks of the whole district are more or less permeated. In many of these veinlets and seams some very high grade chalcocite and cuprite are found, associated with the malachite. The shattered condition of the mineralized rocks conclusively shows that the region has been subjected to secondary movement, and at a time prior to the filling in of the interstices, due to fracture, with the cupriferous solutions.

This latter movement may be accounted for by the intruding of the granite core into the contiguous rocks; the latter, afterward becoming the copper bearing series. It seems that infiltration was downward, and the sutures composing the stock-work of veins and cracks were filled by the cupriferous solutions, leached from the surrounding country rocks. From a cursory examination of the district made on April 5, 1904, by the writer, it appears to have ultimately a reasonably fair future for success.

CHAPTER IX.

VIRGINIA (Shakespeare) AND PYRAMID DISTRICTS.

These districts lie in the pyramid range of mountains immediately south of Lordsburg in Grant county; the locality was known as Ralston (Shakespeare) in the early days. Prospecting began here as far back as 1870, and a veritable boom centered in this region about that time. A man by the name of Ralston was the principal operator and was connected with one of the leading banking concerns in San Francisco, and who, a few years later, committed suicide by drowning in San Francisco bay.

Some sharpers are said to have salted the country in the vicinity of Lee's peak with diamonds in the early '70s, and reaped quite a harvest of wealth from the unsuspecting "tender foot." It is claimed that they had the localities carefully marked where the gems were hidden and pretended to discover them. One diamond is yet said to remain in the sands of that section, since the locality became lost and the tricksters failed to recover it before leaving the country.

An illustration of the excitement that existed in New Mexico, in the boom times on the advent of the Santa Fe railway, is given from a copy of an old "dodger," to which the writer is indebted to Dr. M. M. Crocker, of Lordsburg. It was printed on blue paper 8 x 12 inches and reads as follows :

HO! FOR THE

GOLD AND SILVER MINES

OF

NEW MEXICO.

"Fortune hunters, capitalists, poor men, sickly folks, all "whose hearts are bowed down and would live long, be rich, "healthy and happy, come to our sunny clime and see for "yourselves.

"The Atchison, Topeka and Santa Fe Railroad, has struck "the Rio Grande, and is pushing down the rich valley, flanked "by mountains full of gold and silver ores, sulphurets, "carbonates, chlorides and rich placers not yet prospected.

"Daily mail coaches and telegraph lines to all points. The "whistle of the conquering locomotive will soon be heard in "the newly discovered mining camps of New Placers, Silver "Buttes, Galisteo district and the famous Cerrillos, the "mountains around Albuquerque, the rich leads in the moun- "tains back of Socorro, the mines near Belen, and the mines "near Fort Craig; then comes the world renowned Mesilla "valley with its vines and fruits, encircled by the Organ and "other mountains from which fortunes have been extracted.

": Westward lies Silver City, with its mills and mines; then "comes Shakespeare, the crowning camp of New Mexico, with "San Simon and its Carbonate mountains hard by—the latter "named camp about 4,000 feet above the sea on the divide of "the continent. Here the Rocky mountains end and the Sierra "Madre mountains begin. Here the bold out-croppings tower- "ing fifty feet in the air, bearing gold, silver, copper and lead, "greet the traveler twenty miles distant upon his approach— "the Eighth Wonder of the World. Here the Atchison, "Topeka and Santa Fe and the Southern Pacific railways "have fixed their point of crossing.

"In full view of Shakespeare tower up the Florida, Burro, "Steins peak, Dos Cabezas and Castillita peaks of Old Mexico; "all full of mineral and not yet prospected.

"N. B. Information willingly furnished by all government, "territorial and county officials and citizens generally."

The date of this dodger is not given but it must have been in the year 1880.

In the diffused light of nearly a quarter of a century later, the Eighth Wonder of the World, barely casts a visible shadow and the mighty croppings towering fifty feet in the air have vanished forever from sight.

The old camp of Ralston (Shakespeare), once the scene of frontier activity, is fast passing into decay and many of its fallen walls are playing hide and seek in the shifting sands of the desert.

About three miles to the north of this almost deserted village, the prediction of the "dodger" has been virtually realized; here, two railways cross, creating a new town, which have transformed the early *mirage* on the western plain into a living reality. All hail Lordsburg the Phœnix of a vanished village.

The mineral bearing area in the Virginia and Pyramid districts is about five by fourteen miles in extent.

Eruptive flows and extensive dikes characterize the region. A network of mineralized veins and seams, running in various directions, seem to pervade the districts.

The ore of the two districts is principally a sulphide and carries values in gold, silver, lead and copper. Deep mining would appear to make the region prominent in copper.

Development now going forward on several of the properties will soon demonstrate the value of the ore bodies at depth.

Adjoining the Virginia district on the south is situated the Pyramid district; no definite boundary line exists between the two. From the pyramidal shape of some of the mountains, the name of the district was suggested.

The most important property in the Pyramid district is the Viola group, embracing the Leidendorf mine and mill, belonging to the Pyramid Mining Company.

This mine is purely a silver chloride proposition, and at one time produced a large quantity of silver bullion. For a number of years the property has lain idle and the mill is out of repair.

Another property known as the Silver Tree group, owned by D. E. Miller, is deserving of mention in this district.

In the Virginia district lie the bulk of the locations of the region. At the Aberdeen mine a milling plant has been in operation at intervals for several years; the property has produced considerable values in gold, silver and lead. The mill consists of rolls and steam stamps for crushing, and has two Wilfley tables; its capacity is 15 tons.

The Superior and associated mines of the group are extensively developed and have a large quantity of low grade ore on the dump; gold, silver and copper are the metallic values.

Since nothing more than extensive development is going on in the camp it will suffice to enumerae the principal properties:

Wabash group, Cobra Negra group, Dundee group, Ontario and McGinty, Galena Prince group, Shoo-fly group, The Three Heroes, Carrie, Docotah Pearl group, Eighty-five group, The

Navy group, Ninety-nine group and Lena group; the latter group embraces the Miners Chest.

In connection with the latter property a 75-ton concentrating plant was installed in 1901, consisting of four Huntington mills; four Wilfley, three Standard and one Bartlett tables, which proved a failure. The erection of such a plant at Lorbsburg, five miles away from the mine and transporting the ore by traction engines, when the mine itself was not proved or properly developed, seemed injudicious and ill advised on the part of the management; the ultimate result was no surprise, to any person skilled in successful mine management.

Such disasters have been too often the case in New Mexico; and the practice of erecting plants before the property is fully developed and installing mills and machinery by one unskilled as a practical metallurgist and mining engineer, cannot be criticised too severely. Such calamities are not only visitations of financial distress and ruin on those furnishing the capital, but it inflicts injury, and oftentimes gives a meritorious district a "black eye," from which it takes years to recover.

Gold Hill District.

Gold Hill district is situate twelve miles northeast of Lordsburg and forms another region of Grant county's numerous mining camps.

The center of the district lies in a basin very similar to that of Santa Rita. The core of this basin is a feldspathic granite, and trachyte flows seem to encircle the whole; eruptive dikes break through the porphyry and granite series of rocks, which are most intimately connected with the occurrence of the ore. The veins are all fissures or contact-fissures. Gold and silver are the only values found in the ore; occasionally some copper is found in a few properties.

David Egelston is the pioneer and discoverer of precious metals of the camp. He is a '49er and a typical prospector of the past generation. Egelston came into Gold Hill from Mexico with two partners, Robert Black and Tom Park, in September, 1884. They located the Gold Chief, which they developed during the fall and winter, and the following

year, 1885, sold it to Foster and company, who erected the
Standard mill in 1886.

The Standard group, consisting of the Standard, California,
Golden Chief, Noon-day, Eighty-six and the Little Charlie,
is patented property and supplied the Standard mill with ore;

Fig. 12—DAVID EGELSTON, a '49er and Discoverer of
Gold Hill. Photographed by F. A. Jones, April, 1904.

the principal amount being taken from the Standard and Gold
Chief claims. This mill was operated until refractory ore was
encountered; it has been idle now a number of years.

Another mill was erected about the time the Standard mill
was being operated; this is known as the Dr. Woods mill.

Frank G. Cline, lessee, has been running 5-stamps of this latter mill since 1902 in connection with a small cyanide plant, with marked success. Cline took most of his ore from the Gold Belt lode; he also worked over an old tailing dump by cyaniding, which accumulated from the original operation of the mill, with satisfactory results.

The Little Chief group is owned by the redoubtable David Egelston, who still is enthusiastic over the prospect of making a fortune as he was in the early days of California, over half a century ago.

A number of properties are fairly well developed with good showings of becoming producers, if properly managed. Among that class are the Lottie and Golden Culley, Summit group, Allie and Carrie Lee. The more favorable prospects are Beta and Gamma, Never Fail, Western Belle group, Alma group and the Rattlesnake.

Malone District.

This district is northwest of Gold Hill a few miles in Grant county, and was named after the discoverer John B. Malone, in 1884.

In Thompson's canyon and Gold gulch, tributary to Malone district, placer mining was conducted a number of years prior to 1884. Fred B. Malone, S. J. Wright and John Brown during April, 1904, made some new discoveries in a westerly direction from the old find, about one mile.

Placer gold seems to be plentiful in certain gulches which are now being worked by Lordsburg parties. It is understood that a large quartz and concentrating mill is now in course of erection at that point; the placers would indicate gold lodes in that region.

Eureka (Hachita) District.

About six miles southwest of the new Hachita postoffice, in Grant county, is the old Hachita mining camp.

This camp flourished a short time in the early days, but has remained very quiet since 1890.

Turquoise is mined at old Hachita, and of a very fine quality. These mines were worked to a certain extent by the aborigines and perhaps later by the early Spanish explorers.

Numerous old pits, quaint tools and pottery were in evidence at the time the first American prospectors came into this region.

Most of the veins are contact between limestone and porphyry; the mineral is enclosed in the limestone near the contact.

The principal ore from the district is a silver-lead carbonate. Occasionally copper predominates in some of the locations, as in the Copper Dick group especially; the copper feature is not so strong in the Klondike and in the King group.

In the silver-lead carbonate properties are the American group, Prize, Michigan group, Lady Franklin and Silver Bell.

The production has been very light for the last several years. The ruins of an old smelter serve as a memento of the former glory of the now quiet camp.

Fremont District.

In a southeast direction from the Hachita postoffice, Grant county, the Fremont mining district is situated. All of the mineral deposits lie in or along the contact between lime and porphyry. The ores are principally a silver-lead carbonate; in some instances good values of copper are found associated with the prevailing type of mineral. Zinc, also, seems to occur rather plentifully in a few properties. The Jack Doyle mine is the original and best known property in the district. Next in prominence is the Bee Hive, which is more extensively developed than the Jack Doyle. The American and Sulphide are fairly good prospects. This district, like the Eureka (Old Hachita), has never been a large producer.

Apache No. 2, District.

The naming of this district was due to the frequent raids made by the Apache Indians at the time the region was first being prospected.

In one of those raids, Robert Anderson, the discoverer and present lessee of the Apache group of mines, was severely wounded in an engagement, which resulted in the permanent disabling of one leg.

The Apache group of mines is the principal and only

producing property in the district. It is, in fact, the best property in that section of New Mexico.

The ore bodies lie in the limestone at or near the contact with the porphyry. The ore is essentially a silver-copper carbonate; no gold or lead present. A steady production has been going on for years under the efficient management of the lessee.

This seems to be the only property to speak of in the district; although numerous locations have been made, and partly developed, but without any satisfactory results.

The Apache district is joined on the northeast by the Fremont district and lies a few miles to the southeast of Hachita postoffice.

CHAPTER X.

KIMBALL (Steins Pass) DISTRICT.

Prospecting in western New Mexico, along the Arizona line has never been carried on systematically, until recently. The country generally is much broken and practically destitute of water; timber is also scarce.

From a hasty trip through this region in May, 1903, it would be impossible to give a detailed description of the district.

In the vicinity of the gap, known as Steins Pass, the effect of erosion on the rock system is very much in evidence. The whole country to the north and south for many miles forms one great amphitheater of past volcanic activity. Much of the former roughness of the topography has been smoothed by the agencies of both wind and water and the resulting detritus has formed the flat alkali plains which lie to the east. The country rocks consist of granite, lime and andesite porphyries; the latter, especially, are very noticeable near the Arizona line.

The minerals which abound are gold, silver, copper and lead. The latter mineral predominates to the south at Granite Gap, in the San Simon district, while silver is pronounced at the Volcano mine to the north, in the Kimball district, now under discussion.

Generally speaking, the ores occur in contact veins; although fissures, cutting all formations, are of frequent occurrence.

Most prominent among the properties of this district is the Volcano mine. Two claims, only, form this group—the Volcano and Necessity.

The Volcano vein, apparently, is the mother lode of the district. The strike is approximately north and south and the dip is about 10° toward the east to a depth of 200 feet; after which depth the vein stands vertical.

The vein matter is a quartz gangue and varies in width from two to six feet; it lies along the contact of a felsite and porphyritic andesite. The principal value of the ore is in the silver; the gold value is not so pronounced.

A ten stamp, wet crushing, pan amalgamation mill is on the property; the plant has been idle since 1903. The mine is credited with $110,000 production up to January 1, 1904.

South and adjoining the Volcano property, lies in the order named the Wyman, Fraction, Pa-sh-ly-ky and Bachelor claims; to the north are the Boss, the Queen and others; all of which lie on the Volcano lode.

The old Daley group is noted as the pioneer working of the district; it having produced in the early days.

The Beck group consists of nine claims and is now being extensively developed by the National Gold and Silver Mining Company. Other new companies have been organized during 1903 and expect to begin systematic work at once.

Several prospects are being developed which include the Merrimac, Buckhorn, Dewey, Carbonate Hill, Coon, Volunteer, Mayflower, Horse Shoe, Iron Clad, Wild Eagle, Buckeye, Colorado, Arizona King, Ohio and Gold Quartz.

Steins Pass, now famous in the Peloncillo mountains, was named in honor of one Captain Stein who fought his last, yet victorious battle with the fierce Apaches, in 1873.

This engagement occurred at Doubtful canyon on the line of New Mexico and Arizona, ten miles northwest of the pass at Steins peak. The old Butterfield Stage route lay through Steins Pass (the point where the Southern Pacific railway passes), through Doubtful canyon and through Apache canyon in the Chiricuahua mountains, the next range west, near old Ft. Bowie. Through these three passes, the early emigrant trains to southern California were often ambushed by the Indians.

Captain Stein was said to have been guarding the passes at the time he was killed.

The first prospecting was done in the Kimball district by California and Nevada prospectors about the time the man Ralston (who suicided afterward in San Francisco) created the excitement at Ralston, now the Virginia (Shakespeare) district, which lies immediately south of Lordsburg; this was in the latter part of the seventies of the past century. The second attempt at prospecting was conducted by Williams, Murphy and others who drifted into this region on the crest of the tidal wave of mining excitement, which was passing over the territory. The Apache Indians, at that time, soon forced the aggressive prospectors out of the country.

About the years 1883-9 the first real prospecting was begun by T. R. Brandt, John Corbett, Frank K. Wyman, Chas. R. Smith, Tom Fox, Robert Williams and later Bill Morris.

The three first named prospectors shortly afterward turned their attention to the San Simon district, lying immediately to the south, where they met with success in the Granite Gap property.

San Simon District.

This district is a counterpart of the Kimball district already described. It lies south of Steins Pass, the Southern Pacific railway is the dividing line between it and the Kimball district. The Peloncillo range of mountains embraces both the Kimball and San Simon districts and the general geological characteristics of both districts are practically the same.

The principal mine of the district which is at Granite Gap was successfully developed and operated by Corbett, Wyman and Brandt, who disposed of the mine to S. Chas. Pratt of El Paso in 1903.

The deposit occurs along and in the lime contact and is an ideal fluxing ore; the smelter at El Paso, readily pays the railway freight charges and treats the ore gratis in order to secure it as flux. The ore is a lead carbonate and carries on an average 30 per cent. iron, 12 ozs. silver and 35 per cent. lead, with a lime gangue.

The Johnny Bull lode, Little Lucile, Mineral and Mineral Hill group are the other properties best known; the latter group is owned by the Mineral Mountain Mining Company of Chicago.

California District.

This district was organized and established May, 1904, and is the newest in the territory.

It lies, in part, in southwestern Grant county, and the line between New Mexico and Arizona divides it; the greater portion of the district is in Arizona.

The nearest railway point is Rodeo, a station on the El Paso and Southwestern Railway.

A camp by the name of Paradise has sprung into existence and the ground has been located, so it is said, for nearly twenty miles around.

The cause of the excitement is over a rich copper find in that region; gold, silver and lead also exist.

The first company organized to operate in the district is known as the Chiricuahua Development Company.

Nothing definite is yet known concerning the formations, nature of the veins and ore bodies.

Steeple Rock (Carlisle) and Black Mountain Districts.

In the extreme western part of Grant county, near the Arizona line may be found the old Carlisle district, now known as Steeple Rock. This district was first prospected by Judge Potter in 1881.

The Black mountain district, is in reality, only a particular portion of Steeple Rock.

A few years later much development was carried on and several large companies were organized to develop the region. The Steeple Rock Development Company and the Laura Consolidated Company were the most prominent; these organizations were controlled by English capital.

A 60-stamp mill was erected by the first named company, with all necessary accessories, and operated with a certain degree of success, until the heavy sulphide ores were reached. In 1897, work was suspended and the plant dismantled and most of it moved to various parts of the country; it being the intention of the owners at that time to erect a different kind of plant, using a different method in the ore treatment. The new plant, as yet, has failed to materialize.

The ore is a bluish-white quartz and very hard, carrying values in gold and silver; this class of ore is the prevailing type over the entire district. Frequently, sulphides of lead and copper are associated in the vein filling.

Most generally the veins are of the true fissure type, existing between walls of porphyry; the character of the porphyry is, perhaps, an augite-andesite.

At present but little is going on in the district and the production has decreased to almost nothing.

The principal groups and locations are as follows:

Carlisle group, East group, Jim Crow group, Laura group, Big Horse Shoe group, National Bank group, Bill-ali group, Summit group, New Year's Gift group, Big Four group, the Alabama, Henrietta, Hortense and the Little Mack group.

In the latter group the vein is a contact between porphyry and lime, and the ore is characterized as bornite, carrying some values in gold and silver.

Anderson District.

Lying in Grant county, about midway on a line connecting Silver City and Steeple Rock, and on the east side of the Gila river, is the locality known as the Anderson mining district.

This particular region was first exploited by Lou Anderson, in 1881, who made several locations about that time.

Anderson was killed in 1884, either accidentally, by the Indians or by suicide, it was never known definitely how; he was found a few hours afterward lying dead on his face with a bullet hole in his head. His winchester was by him with an empty shell in the barrel; the fatality occurred near his mining prospects.

The Anderson district has never been a producer; the chief reason being its great distance from transportation.

A number of copper prospects are held by the Alessandro Copper Mining Company, as also, a large fluorite lode; this latter deposit is known locally as the "onyx property," and is situated immediately on the east side of the river at the mouth of the box.

To the southwest of this fluor spar deposit about three and one-half miles is the celebrated ricolite quarry, which is mentioned under the chapter on building and ornamental stones in this volume.

Granites, green-stones and intrusive dikes characterize the rock formation of the district.

It would seem from a cursory examination of the district, that a considerable production would be the outcome, were transportation facilities more favorable.

Telegraph District.

This is another one of Grant county's numerous mining districts. It lies to the northwest about six miles, and on the opposite side of the river from the Anderson district.

The district is practically inaccessible; the mountains are very rugged and can only be crossed by trail.

From the point of the old Telegraph Mining Company's mill (now Dr. H. W. Brown's mill), south to the southwest corner of the Anderson district, the Gila river is completely boxed. The river at one time evidently was dammed here for a distance of seven miles, due to the stupendous volcanic disturbances which engulfed the entire region in this part of New Mexico. These disturbances appear to have taken place about the close of the Tertiary.

Telegraph district was discovered in July, 1881, by A. J. Kirby, a Texan; the first claim located by him was the Tecumseh lode, a silver property. Del Potter and Dorsey brothers were among the first arrivals, immediately after Kirby. Much excitement existed throughout the Territory about this time and the country was crazed over the numerous discoveries and mines of high grade silver ores.

About three years after Kirby's discovery the Telegraph Mining Company was organized, and in 1885 it erected a 15-stamp mill on the Gila, at the head of the box in the river. This enterprise was first attended with marked success; the ore body, however, being small was soon exhausted and the concern collapsed. The mill was afterward removed from the district.

Dr. H. W. Brown and associates of Silver City, erected in 1903, a small leaching plant with a capacity of about five tons a day, on the old Telegraph company's mill-site. Dr. Brown's method of recovery of the silver values is similar to that of the Russell process.

The ores of the district are usually composed of an indurated bluish quartz, containing argentite with occasional cerargyrite.

The vein in which Dr. Brown is mining his ore is a true fissure in granite; the silver values seem to have been leached from former superimposed strata, collecting in the fissure receptacle of granite. The vein from which the old Telegraph company extracted its principal ore, lies about one-fourth of a mile to the south of Dr. Brown's claim, near the top of the mountain and is a contact between porphyry and shale.

It is said that the district took the name Telegraph, due to a remark made by the discoverer of the first rich ore, who said he was going to climb to the top of a certain high mountain (now known as Telegraph peak) and from there would telegraph the news around the world of the greatness of his find.

CHAPTER XI.

ORGAN DISTRICT.

Lying to the northeast of Las Cruces, in Doña Ana county some fifteen miles, is the Organ mining district.

Rising abruptly from the level plain the jagged Organ peaks present an appearance similar to the pipes of an organ, from which the mountain range took its name.

A conspicuous isolated granite peak, similar in character to the sharp cones or spines of the Organ mountains, marks the southern terminus of the San Andreas range: this lone spine is known as the San Augustine peak. At the foot of the south slope of this monolith is the San Augustine pass, which separates the Organ mountains from the San Andreas range. This famous gap in the mountains affords a splendid wagon road and renders Gold Camp (Black mountain), which belongs to the Organ district, accessible from the west side of the range.

Lying upon the granite, or its gneissic equivalent, is a quartzite that seems to have been derived from a schistose-quartz which was previously a superficial portion of the subjacent granitic series. The age of both the quartzite and granite is a question which has not been definitely determined. Conformably to the quartzite is a partially altered silicios series which is immediately followed by the massive Carboniferous limestones. It is in or near these limestones in which the ore bodies of the district lie; deposition being chiefly due to pneumatolytic action resulting from contact metamorphism. The commercial metals found in the district are those of lead, silver and copper on the west side of the range; while gold is the principal metal on the east side.

The property deserving first mention on account of its prominence and early history is the celebrated Stephenson-Bennett lead-silver mines. The Stephenson lode was discovered by a Mexican in 1849. Hugh Stephenson living near-by on the Rio Grande, soon afterward became a partner and

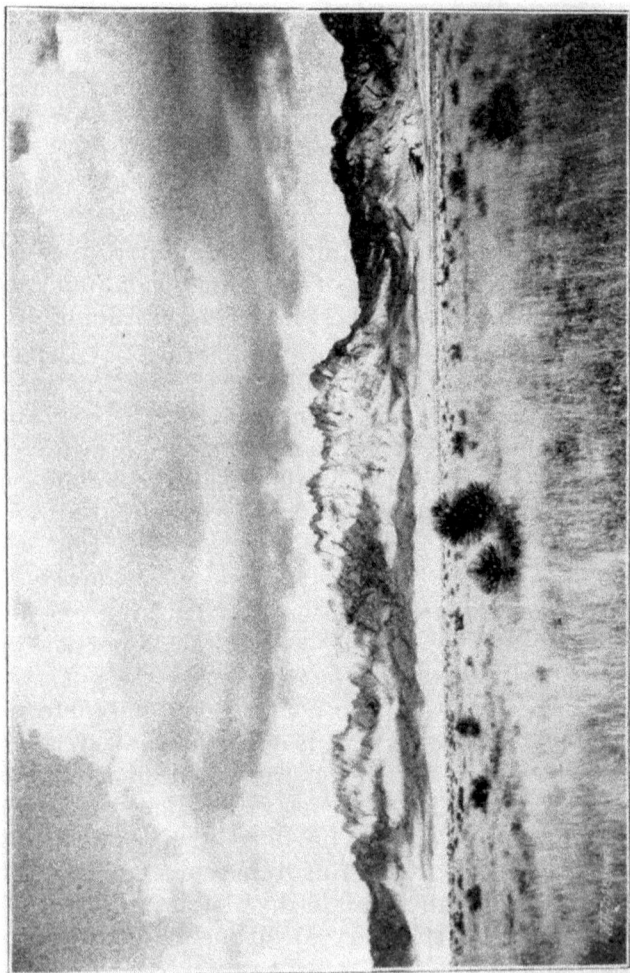

Fig. 13 - ORGAN MOUNTAINS. Photographed by F. A. Jones, 1899.

later on the sole owner of the mine.   In 1858 he sold the mine
to army officers of the United States, who were then stationed
at Fort Filmore, for $12,500.

In the early working of this mine no powder was used and
what the pick and shovel failed to do was left undone.   During the years 1854-7 work was carried on in a more systematic manner than was formerly done; and the estimated production during that time was between $80,000 and $90,000.
The ore up to 1882 was carried out on the backs of the laborers, as no hoist or windlass had ever been erected up to that
time.   Smelting operations were conducted in an adobe
furnace, on the Rio Grande at Fort Filmore, sixteen miles
away; the ore being transported on the backs of burros.   An
ordinary blacksmith's bellows supplied the blast.

In the Mineral Resources of the U. S. for 1870 on page 412,
the property is briefly mentioned as a producer.

Work was hampered in the earlier years by the Apache
Indians and also, by the Mexicans whose feelings were bitter
toward Americans, as a result of the Mexican war.

The ore first mined came from a parallel ledge above the
present main workings; some of which carried astonishingly
high values in silver.

The lower vein is opened by a tunnel some 200 feet long,
driven into the mountain side at right angles to the veins.
At the east end of the tunnel where the same cuts the lode a
winze is sunk on the vein to a depth of 200 feet, with levels
running either way at each hundred feet.   The main vein
dips to the west about 80°, and the strike is approximately
north and south.   Some splendid ore bodies are opened up
exposing extensive chambers of high grade sulphides and
carbonates of lead-silver.   Mud caves are frequently encountered and near these rich ore bodies are usually found.

Most all the properties of this camp are good illustrations
of mineralization due to the effect of contact metamorphism.

The hanging wall is a hard blue limestone and the foot wall
is a granite-porphyry dike; the contact is very strongly
marked and can be traced for a distance of nearly eight miles
along the foot of the range.   The ore bodies of the Bennett
claim are principally in lime, yet the porphyry intrusive
frequently breaks through the limestones to the fractured

Fig. 14—STEPHENSON-BENNETT MINE.

zone or fault-fissure in the lime rock, and thus the actual contact of the dike and the vein occurs in a number of places. It
was observed that cavities in the limestone not having communication with the dike are destitute of any minerals. Therefore, it would appear that the dike is the source from whence
the metallic values came.

No complex system of faulting has been found in the district; although dislocations of minor importance are frequently encountered.

The production of the Stephenson-Bennett mines from
their discovery to the year 1904, is approximately $500,000;
of this amount $200,000 is accredited to the decade between
the years 1890 and 1900.

The principal minerals found in this property are galena,
argentite, anglesite and wulfenite. Some of the finest
crystals of wulfenite ever found in New Mexico came from
the Bennett lode.

About six miles to the south of the Stephenson-Bennett
property is the Modoc mine; this also, is a lead-silver proposition and apparently lies on the same contact. This property
was located in 1879, and has been worked in a spasmodic way
ever since.

The chief developments consist in an incline shaft 185 feet,
which follows the dip of the vein. An adit tunnel 85 feet in
length intercepts this shaft at 95 feet depth on the incline.
This tunnel is extended to a distance of 85 feet beyond the
shaft. The vein is about seven feet in width; the grade of
ore, on an average, is rather low. A dry concentrating plant--
the Hooper pneumatic process—was installed in 1902, but
was not a success. Hoisting machinery, compressed air for
drilling and a rope tramway constitute the principal improvements. The property seems to be one of considerable merit,
but has been unfortunate in its management; it is, at present
1904, idle.

A number of other lead-silver claims, lying between the
Modoc and the Organ postoffice, cover this main contact
fissure; some of which may eventually become producers.

No mining claim in the Organ district has made such a
favorable showing as the Torpedo, in the same length of
time. It lies about 200 yards east of the Organ post office on

the main contact of the district. Four years ago it was only a prospect; now, in 1904, it is valued at a quarter of a million dollars and has over 3,000 feet of shafts and drifts. The property is copper-silver; its ores are oxides, carbonates, sulphides and silicates of copper. The silver is chiefly argentite. The dimensions of the ore deposit are of much magnitude and the deposit itself is similar in many respects to those of Bisbee, Arizona.

It is a contact deposit between porphyry and limestone; the former constitutes the foot wall, the latter formation the hanging wall. The strike of the lode is northeasterly and the dip is slightly toward the northwest. Great trouble is experienced in handling the water; the flow is very strong and all the property to the north along this contact is apparently drained by the pumps in the Torpedo mine.

The dike at this point is somewhat different in its constitution than at the Stephenson-Bennett and Modoc properties, which lie to the south. Here, the mineralizing influence of the dike has supplanted the lead for copper, but retained the silver. That mineralization was induced by contact metamorphism there is no room for doubt.

No evidence has been observed in the district which would indicate metalliferous deposits, prior to the time of the dike.

The effect of metamorphism is better studied on the Copper Bar and Memphis properties, which adjoin the Torpedo on the north, than at any other place in the district.

The character of the limestone has been entirely changed on the Copper Bar and Memphis, which has resulted in the formation of massive garnet and epidote. This alteration is more especially noted on the Memphis, and bears a close resemblance to the copper districts of both Clifton and Morenci.

The Memphis and Copper Bar are not entensively developed, but they are most favorably situated and are to be regarded as valuable copper properties.

One mile farther to the north is the Excelsior which has 500 feet of work done on it; a very good showing of copper is noted. The Merrimac and Little Buck are adjoining claims and lie to the northeast of the Excelsior about one mile. The former is a lead-zinc proposition; the latter is silver-gold. The Little Buck deposit has been an anomaly; some $50,000 in silver and gold have been taken out within a few feet of the surface. Apparently this was a pocket between the dolomite and porphyry, the values having been leached from the porphyry and thus concentrated by descending waters.

Other properties in this portion of the district are looked upon with favor, although they can be classed nothing more than prospects.

Gold Camp (Black Mountain), which lies on the east side of the mountain range, some ten or twelve miles from the Organ postoffice, belongs to the Organ mining district also. This section was first prospected about 1883 by Pat Breen, John and Henry Foy and others. The geology of this camp is entirely different from that of the Organ side; since it lies at or near the great fault line which runs along the east side of the range, extending from near El Paso, north to the northern terminus of the San Andreas.

Granitic and metamorphic rocks are traversed at intervals by quartz-porphyry dikes. These dikes have influenced the segregation of gold along the planes of contact. Some copper and silver are also found in association with the gold. Taken as a whole the region of Black mountain may be considered one of low grade gold ore.

The Mountain Chief was the first property located and possesses considerable merit. The most extensively developed property is the Mormon mine; the Dona Dora is also being developed and looks well. Lying between Black Mountain and Mineral Hill is the Oriental lode. In Texas Canyon to the south is the Mascot and other claims. To the north of Gold Camp about eight miles is Bear Canyon, which contains some promising lead prospects. Some of these lead prospects are high up in the mountain and would need an aerial tramway to properly handle the ore. The Pharmacy and Dona Dora are prospects which would deserve notice, as they are accessible and lie centrally in Gold Camp.

It seems that plenty of water can be developed in this region; transportation being the greatest drawback.

West of the Organ postoffice about twelve miles and at an equal distance north of Las Cruces are situated the Doña Ana mountains. Some prospecting has been carried on at intervals in this little group of isolated peaks, but nothing of value has yet been found. Although many surface indications of minerals may be seen at various points. It would, therefore, not be surprising to hear of some bona fide discoveries made in that section at any time.

Hembrillo District.

The Hembrillo district lies in the southern part of the San Andreas range, north of Organ in the northeastern part of Doña Ana county; to the east are the gypsum hills in Otero county.

This section of the San Andreas mountains is identically as the region further to the north as described under the chapter of the districts of the Sierra Oscura, and San Andreas ranges. In a majority of cases the veins are contacts, lying between lime and porphyry.

Copper glance appears to be the principal ore of this district; quite often fair values in gold, silver and lead are encountered. Really but little development has ever been done and it is not possible yet to determine the worth of this region.

The Base group, a lead property and belonging to the New Mexico Lead Company is the best known and developed. Another group known as the Little Monte is being developed by Capt. Thomas Brannigan of Las Cruces; this property is copper and gold.

The Planet Mars and two other groups are copper bearing and belong to Major Llewellyn, Governor Otero and others

CHAPTER XII.

LAS ANIMAS (Hillsboro) DISTRICT.

It was on the 20th day of April in 1877 that gold was practically discovered at Hillsboro by Dan Dugan and Dave Stitzel,
who were prospecting at that time on the east side of the
Mimbres range. When crossing over what is now the Opportunity mine on that day in the Las Animas district, some
float was picked up which Dugan pronounced "no good," saying "anyone ought to know that such a formation as this
carries no mineral." Stitzel, notwithstanding, put a few
small pieces in his pocket and carried them to an old quartz
mill on the Mimbres river where he had them assayed. To
the great surprise of the two prospectors, the ore ran $160.00
in gold per ton.

About one month later on May 13th, they came back and
located the Opportunity and Ready Pay mines.

Under adverse circumstances they managed to mine and
hauled over five tons of ore to the Mimbres mill, which netted
them $400. In August of the same year, the first house was
built and the town of Hillsboro, now in Sierra county, started.

The famous Rattlesnake (more commonly called the Snake)
mine was discovered by Frank Pitcher and Dan Dugan in
June, 1877. This discovery was accidental, and was made by
those two prospectors when returning to their camp in Ready
Pay gulch, having sat down under a clump of oak bushes to
rest, striking and breaking loose rock which lay about them
as they talked, free gold was found in some of the broken
fragments. Owing to the fact that they had just killed a
large rattlesnake while resting at that spot, it was decided to
name the claim the "Rattlesnake" lode.

The news of the new gold district soon spread, and by fall
quite a population had gathered about Hillsboro.

In November, 1877, placer gold was first found in the district by Hank Dorsey in the Snake and Wicks gulches. Adjoining gulches were afterward prospected and the discov-

eries soon spread and embraced the rich diggings on and about Slap-jack hill.

During the winter of 1877-78, a mine operator by the name George Wells turned into the stores and saloons of Hillsboro $90,000 in gold dust and nuggets which he had taken from Wicks gulch.

The first ore taken out of the camp, excepting the five tons that were hauled to the Mimbres mill, was worked in arrastras, built in what is now Mattie Avenue, Hillsboro, in front of the Stage stable, at the end of the Stage line now operating between Lake Valley and Hillsboro.

A Mr. Fresh built two steam arrastras in the fall of 1877; in the fall of the following year, and in partnership with a Mr. Wicks, the two erected a 10-stamp mill on the old arrastra site, which may be seen at the present time.

The country around Hillsboro was, in the early days, a part of Socorro county; later, all land situated within a radius of two and one-half miles of Hillsboro was attached to Doña Ana county. In April 1884, Sierra county was organized by an Act of the Legislature in 1883, the preceding year, and Hillsboro was made, and has since been, the county seat.

Of the old timers Dan Dugan died in 1884; Dave Stitzel is living in Hillsboro, and whom the writer had the pleasure of meeting; Hank Dorsey resides in Silver City and is now seventy years old, and who still expects to make another fortune in mining; George Wells has drifted away from the scenes of his early mining operations and is lost to the now few, old time prospectors, who yet remain in New Mexico.

Such is the brief history of the discovery of gold in the Las Animas mining district at Hillsboro.

Andesites trachytes and intrusive diorites, seem to be the principal eruptive rocks of the district; although a high, narrow, isolated ridge, separating the town on the north from the mines, is covered with basalt as well as some other patches, as observed in the vicinity of the Opportunity mine.

Much birds-eye porphyry is found throughout the entire district and the mineralization seems due to some sort of connection with the same. Most of the veins are well defined and at times attain considerable width. The general strike of the lodes seems to converge in the direction of Las Animas peak. All of the principal properties, with but few exceptions, are located on a ridge covering the parallel or radial veins as heretofore alluded.

Beginning at the west end of the ridge the principal claims occur in the following order, viz:

Golden Era, Empire, Garfield and Butler, Richmond, Eldorado, Bonanza, Morning Star, Snake, Moccasin, Opportunity, Ready Pay and the Wicks. The rich placers of Wicks gulch, and around Slap-jack hill evidently resulted from erosion of the lodes of the above mentioned claims. The erosion must have been very great, as the present topography of the country would indicate, and the enormous beds of gravel and debris in the direction of the Rio Grande would suggest.

Beside gold, there are found some copper and lead ores carrying good values; also, the beautiful mineral endlichite is found on the eastern slope of the district.

This latter deposit of rare mineral is in a contact-fissure, with a lime foot wall and a shale hanging wall. The contact is irregular, though very strong, extending a distance of nearly 4,000 feet.

It is said that this is the largest body of Vanadium ore known in the world. The property is known as the S. J. Macy lode.

The Prosper, Cincinnati, Mascot, Mining Company, Lupey, Summit, American, Virginia, Sherman group, Rubicon group, Whaleback and Perche groups, Eureka, Bob-tail, Bull-of-the-Woods, Catherine, Lilliput, Prince Henry, etc., are other properties that are being developed. A number of placer properties are being worked in a small way, generally by Mexicans, which produce on an average of $450 per month.

Some seven or eight miles further to the north and west is the Andrews postoffice, where the Golden Rule group of mines is located, and which has had considerable production. Nearby are the Chance group, Emperor group, Little Nell group, '97 and '98 mines.

The estimated production of the camp to January 1, 1904, including both placer and lode claims is placed at $6,750,000, principally gold.

Pittsburg District.

This district lies immediately on the east side of the Rio Grande and embraces the Sierra de los Caballos, in Sierra county, northwest of Rincon.

For many years the attention of prospectors and capitalists has been periodically attracted to the district, chiefly due to the existence of lead and copper found in many localities of the range.

Not until recently did the district come prominently before the public, when a great rush to the newly discovered placer fields was made.

Some two years prior to the knowledge of the public of these gold fields, a Mexican by the name of Encarnacion Silva, made periodical visits to Hillsboro, some twenty-five miles away, and disposed of his gold dust and nuggets, and who persistently refused to disclose the locality from whence his source of wealth came.

On one visit, however, to Hillsboro, this cunning native, due to an over indulgence in intoxicants, became talkative and the secret of his find became known; on this latter occasion he was accompanied by his cousin, Bernardo Silva, whom he had taken into his confidence. This was on Sunday night, November 22, 1903, when several persons started at once and rode through to the diggings in the night and were on the ground making locations early the following morning. It was not until this same morning of the 23rd that the news became generally known in Hillsboro, when a general stampede and exodus of the population of the village resulted. Everything partook of the nature of an old fashion mining boom, similar to that in the early days.

Excitement ran very high and it was claimed that the District Court, then in session at Hillsboro, adjourned and left for the golden Eldorado.

The writer was on the ground, the middle part of the following week representing Albuquerque men, and on Saturday night December 8th, by the flame of the camp fire, was unanimously elected chairman to preside over the deliberations of that unique body of fortune hunters. From a spectacular standpoint, in which winchesters and six shooters were in evidence, some of the crowd standing and others seated on boxes and empty beer kegs, the proceedings were rendered most impressive.   Never was a more orderly or harmonious meeting held; every motion put and seconded was unanimously carried with a Stentorian shout coming from a thousand throats "I," making the very mountains reverberate with the

Fig. 15—BLOWING GOLD FROM SAND. Photographed by F. A. Jones, December, 1903.

(The man to the right in the picture, who is looking up, is Encarnacion Silva, the discoverer of gold in Apache canyon.)

sound, and which was echoed back in turn to the great valley of the Rio Grande from whence it came.

The placers seem to be confined to or near the Apache canyon and to the northern branches of the same.   Only two gulches to the present time, May 1, 1904, have been found

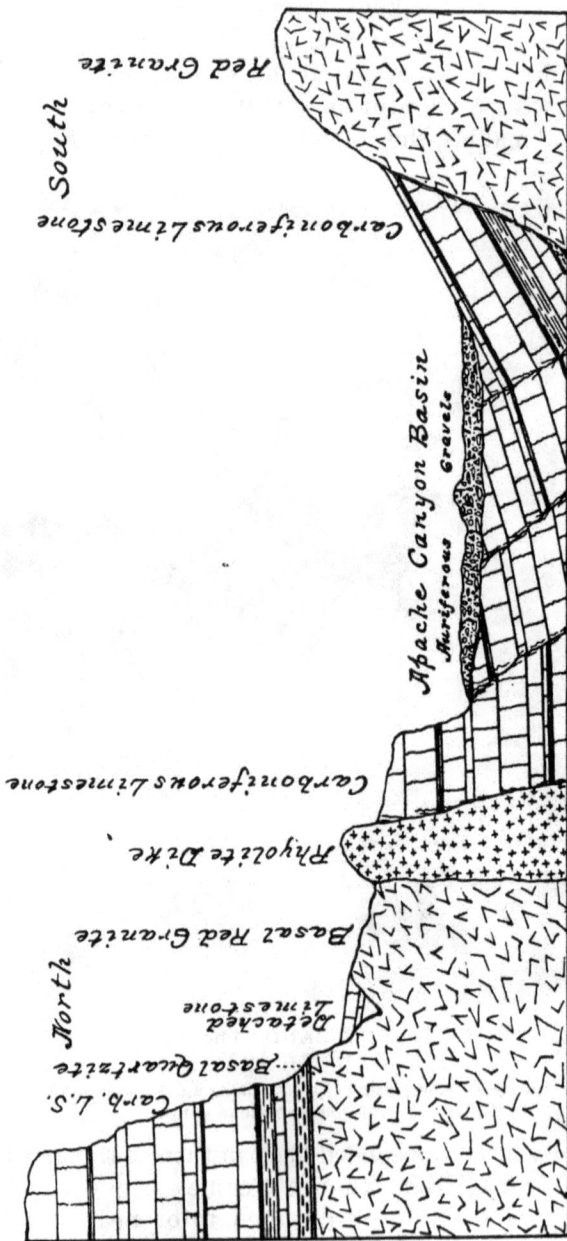

Fig. 16—SECTION AT RIGHT ANGLES ACROSS APACHE CANYON, IN APACHE BASIN, ABOUT TWO MILES
BACK AND NEARLY PARALLEL WITH THE RIO GRANDE.

where gold exists in paying quantities; these are known as the Silva and Trujillo gulches, respectively. The former gulch is where Silva originally discovered the gold; the latter gulch was found to contain gold at the beginning of the excitement.

The area of the pay gravel seems rather circumscribed, although development may materially enlarge the field.

All the gold won from the sand and gravel at the time the writer visited the district, was done by means of blowing through a pipe-stem or small tube in the loose sand which covered bed rock, exposing the nuggets and particles of gold, that were afterward picked up with the fingers, or a small stick, moistened with the mouth at the end.

During the following spring a man by the name of H. O. Clark, from San Francisco, introduced a new gold washer which seems to be successful, handling about thirty tons of gravel per day. The new machine consists of a cylinder, three feet long and eighteen inches in diameter, surrounded by a revolving screen. There is a half circle bottom containing riffles about one-half inch deep and one inch in width in which the dirt drops after passing through the screen. The gravel that falls into the riffles is constantly agitated by steel fingers or wormers that separate the gold from the dirt. The values, it is said, are saved so closely that expert panners are unable to produce any colors from the tailings.

A transverse fault nearly at right angles to the profound fault which formed the Rio Grande basin, seems to have been due to a fissure from which a great flow or dike of rhyolite had its egress and which gave Apache canyon its present trend.

Intimately connected with this rhyolite member at and along its contact, appears to be the source of the gold.

This disturbance must have taken place at the beginning of the pleistocene, since the Tertiary gravel beds are tilted and have the same slope as the older series of rocks in the canyon; the throw must have been 2,000.

It is evident that the gold did not come far as it is quite flaky; in one instance the writer saw a flake that had been rolled into a cornet, by some natural process.

The Marion mine is perhaps the most noted lode claim; it

lies in the north end of the district.   This property is a copper proposition and has been a producer for several years. It is opened by a tunnel 700 feet long; the ore is chalcopyrite and chalcocite.   To the north end of the district are situated the Rosa Lee and the Napoleon claims, two parallel lodes, which are quite promising.

Their veins have a strike approximately east and west and stand almost perpendicular.   They are true fissures and break through the overlying carboniferous limestones; the gangue is a white quartz in which the ore occurs.   Some beautiful cubical crystals of wulfenite are associated with the ore and adhered to the walls. The Washington group, Othelia and Iola are favorable copper prospects having a lime contact and are only partially developed.

At the south end of the district a short distance north of Rincon are some deposits of manganese.

The ore occurs as psilomelane, and is of very high grade; the extent of the deposit has never been fully exploited.

Some coal measures exist at and beyond the northeast end of the district, which have never been developed; much faulting seems to have occurred, which would render the beds expensive to work, provided the vein is sufficiently thick.

The mineralization of the Sierra de los Caballos, extends into the Sierra Fra Cristobal range immediately to the north, where but little prospecting in this latter range has ever been done.

Iron Reef District.

This district lies on the west side of the Rio Grande in the vicinity of Palomas hot springs, in Sierra county.

The two principal properties here are the Iron Reef and Meridian.

Very little is doing in the district at the present writing. The ore is principally lead-silver; some gold is also found in the associated ores of the district.

CHAPTER XIII.

LAKE VALLEY DISTRICT.

One of the most remarkable bodies of silver ore ever encountered in New Mexico, was the Lake Valley deposit.

In fact, the Bridal Chamber ore body has never been equaled in richness by any silver mine in the world.

The Lake Valley mines were discovered by George W. Lufkin, a cowboy prospector, in August, 1878, who at that time had headquarters at Hillsboro.

The discovery was purely accidental; the cowboy in round of duty, got off his horse to tighten the girth of his saddle and noticed a peculiar piece of stone, which he picked up and was surprised at its heft. He had it assayed on suspicion that it might be ore.

To his great astonishment, the piece of float ran several thousand ounces of silver to the ton. This piece of float was found near where the Bridal Chamber was afterwards opened up.

George W. Lufkin took in a partner by the name of Chris Watson; these two gentleman (both now dead) realized but little out of their find.

It is claimed that Lufkin received only $10.50 for his share; the property soon passing into other hands.

The fame of this rich discovery soon spread and a rush for the new strike was made by every class of people.

After some development was done on different locations, three companies absorbed all the best and promising claims.

Those companies were the Sierra Grande, capitalized at $2,000,000; the Sierra Bella, capitalized at $1,000,000; and the Sierra Apache, capitalized at $1,000,000; and all composed principally of Philadelphia capitalists.

Operations of the three companies were conducted for several years under the management of the Sierra Grande Company and it was under this management that the famous Bridal Chamber was found. A blacksmith by the name of

John Leavitt (now dead) who had a lease on that particular property, discovered the Bridal Chamber and sold out his lease to the company for a few thousand dollars.

It was on the very day of the discovery of this remarkable silver deposit—the Bridal Chamber—that the General Manager George Daly was killed by the Apache Indians about six miles out from camp in the early eighties.

Dr. F. M. Endlich, after whom the rare mineral Endlichite took its name, first exploited this property. The managers of the property under the old organization, after the death of George Daly, were: Walter A. Hadley, Ellis Clark, Robert Eastburn and Henry Schmidt

Professor Ellis Clark who had charge of the mines a number of years, wrote a valuable paper on this famous property which was read at the Virginia Beach meeting, February, 1894, before the American Institute of Mining Engineers.

The Lake Valley district contains three characteristic eruptive formations, viz: Rhyolite, porphyrite and a horn-blende-andesite. These three eruptives are within a few hundred feet of where the principal ore bodies were found. In fact, the porphyrite lay immediately on much of the rich deposits, including the Bridal Chamber.

The rhyolitic flow is perhaps the latest type of eruptives in the district, and composes what is locally known as porphyry hill which lies to the southwest of the main works.

Monument Peak which is due east of the mines is composed of andesite. The total approximated thickness of the sedimentary formations at the Lake Valley mines is about 1,000 feet.

The lowest of the series is composed of quartzite and limestone and regarded as Silurian. Then comes black and green shales, nodular limestone, blue limestone and crinoidal limestone: this latter limestone is about 200 feet in thickness and represents nearly half the thickness of the lower Carboniferous at this point. The blue limestone is the receptacle in which the ore bodies have occurred.

The ore bodies are something similar to those of Leadville on account of their peculiarity of deposit in the nearly horizontal formations of limestone. Yet, the deposits can not be termed of blanket form, since they occur in beds or troughs which are due to the erosion of the blue limestone. The overhead or hanging wall formation is either a porphyrite or the crinoidal limestone.

The quartzite of the district extends north into the Hillsboro mining camp, about eighteen miles distant.

There have been several theories propounded concerning the origin of this ore; but it is more than probable that the leaching out of the metallic sulphides and chlorides of silver from the porphyritic sheet is the true source.

The ore is usually accompanied with quite a percentage of manganese and iron; and only occasionally is galena present.

A variety of silver and other ores have been found in these mines; such as cerargyrite, stephanite, descloizite, vanadinite, endlichite, proustite, etc. Magnificent specimens of pyrolusite showing the crystallization have also been taken out.

Approximately, the different workings have yielded:

	Ozs. of Silver
Bridal Chamber	2,500,000
Thirty Stope	1,000,000
Emporia Incline	200,000
Bunk-house	300,000
Bella Chute	500,000
Twenty-five Cut	200,000
Apache and all others	300,000
Total	5,000,000

These mines were operated for a period of about fifteen years by the Sierra Grande Company, closing down in August, 1893.

In April, 1900, the the entire property was sold at a private sale to L. G. Fisher of New York City, who had been associated with the original companies and who organized in 1901 the Lake Valley Mines Company.

New development began a little later on under the direction of John Hays Hammond, consulting engineer, with E. H. Bickford, general manager.

Considerable ore has been taken out and shipped under this new organization. The ore lies well down in the blue limestone, as was observed recently by the writer, and is of a mangano-ferruginous character, running low in silver values. This region is limited in extent, in so far as its mineralization is concerned. Every part of the district adjacent to the principal Lake Valley mines is covered by locations and the annual assessments regularly done.

A few of the more favorable locations are the Stone Cabin, Miles Standish, Teddy group and the Centennial State group.

Macho District.

To the south of Lake Valley district in Sierra county, on and near Macho creek, several prospects exist in what is known as the Macho district.

Lead and silver are the principal minerals; although, it is probable that deposits of manganese of considerable importance may be developed. The geological characteristics here are much the same as at Lake Valley. The underlying limestones are entirely covered by vast sheets of andesites and trachytes, through which the principal veins outcrop. Abundance of water exists near the surface which can be developed for mining purposes. The Dude mine is more developed than the others at this point. The vein is a true fissure with a gangue of flinty quartz which carries sulphides and some carbonates of lead, with a little silver and gold.

Lying immediately across the draw to the north of the Dude is the Hudson group of lead properties. This latter property is, indeed, very promising and by proper development would make a good producer. Cerussite predominates, which would point to the association of this mineral with the lime formations below. Many beautiful crystals of wulfenite are mingled with the vein matter.

About eight hundred feet to the west of the Dude is the Jim Crow manganese location, which was made in April, 1903, by S. W. Sanders, and the writer, F. A. Jones. Very little work has been done on the property, yet the indications point to quite a deposit of manganese when properly developed; the ore is pyrolusite, associated with calc-spar. The vein breaks through a decomposed andesite covering.

Bromide No. 1. (Tierra Blanca) District.

This mining section of Sierra county, New Mexico, took its name from the white capped hills or low mountains which are covered by a rhyolite flow, signifying white earth; it lies to the northwest of Lake Valley about fifteen miles, in the foot hills on the eastern slope of the Mimbres range of mountains.

The name of the district proper was given it due to some high grade silver bromide found there in the early eighties of the past century.

Fig. 17--LOG CABIN MINE, photographed by F. A. Jones, Dec., 1903.

The ore and its occurrence here is a counterpart of what is found about Kingston, some ten miles to the northwest.

High grade chlorides and sulphides of silver, with frequent occurrence of native silver, found in and near the contact of lime and prophyry or lime and shale, represent the prevailing characteristic mineralization of the district.

Occasionally pockets carrying phenomenally high values in gold are encountered; these appear to be always near or at the surface, in association with quartzite and porphyry.

This latter feature was the case at the Log Cabin mine where many thousands of dollars in gold were taken out at the grass roots, not exceeding a depth of ten feet.

The remarks applying to the camp of Kingston, concerning the geology of the ore deposits there will in a general way apply here.

Near the head of Trujillo creek is the most prominent producing mine of the district—the Lookout. The lode is a contact between limestone and granite-porphyry and carries sylvanite.

Some shipments have proven to be phenomenally rich in gold and silver values. Most of this high grade ore came from a comparatively superficial depth. The property belongs to two of the old timers, Col. Parker and J. M. Webster.

The Log Cabin mine is also, one of the best known properties in Sierra county and is at present producing; although, not however, to the extent that it formerly did.

Many other properties of the district are fairly well developed; among which the Tierra Blanca group, the Midnight and the Bell are the most prominent.

CHAPTER XIV.

BLACK RANGE Nos. 1 AND 2, DISTRICTS.

All that portion of the country on the east slope of the Black Range of mountains lying between Kingston on the south and Grafton to the north, was in the early days denominated as the Black Range mining district. This large area, fifty miles north and south and from ten to twenty miles east and west, embraced a number of mining camps and districts; all of which are now usually spoken of as the Black Range.

The term "Black Range" was adopted by the early explorers from the fact that these lofty mountains, when viewed from a distance, presented a very dark or black appearance, due to the heavy growth of piñon and pine timber which covers the surface.

Many traditions, superstitions and perils pertaining to pioneer life in the great southwest, are historically interwoven in a woof inseparable from the black band which impresses its inky form in the distant horizon.

Kingston Camp.

The excitement about Hillsboro in its early days led to the discovery of silver in the Kingston camp. There were two parties of prospectors, who first entered the Black Range district and who accidentally met at the present site of Kingston; this was in the latter part of October, 1880. Bob Forbes, Frank Pitcher, Dan Cameron, Jim W. Wilson and H. W. Elliott constituted the first party; the second party was composed of Messrs. Chapman, Heard and Phillips. To the honor of the second party mentioned, belongs the discovery of silver in the Black Range district at the Kingston camp.

The Iron King and Empire were the two first locations and were made by Phillips and Elliott; they afterward cast lots to determine which claim each should take. Phillips drew the Empire, consequently the Iron King fell to Elliott.

Next located was the Eclipse by the Phillips party; then the Brush Heap by a late arrival having the name of Johnson, and afterward the Blackeyed Susan by Forbes and Elliott.

Dan Dugan, famous as one of the original discoverers of gold at Hillsboro, with three partners located the Lady Franklin, the Gray Horse and others. In the spring of 1881 the mining district was organized and named the Black Range.

Several of the old timers who landed in the camp about the time the district was organized, still reside there. Col. John Logan and Col. A. W. Harris in the spring of 1881 (the latter still a resident of the camp) have the distinction of being the first persons who rode into Kingston in a vehicle (an old ambulance); they coming from Hillsboro, found it necessary to cut away the brush and trees at numerous places along the trail in order that their ambulance might pass.

It was at Kingston that the prospector sallied forth a poor man in the morning, returning a millionaire in the afternoon for he had "struck it rich". Here fortunes grew in a single day to vanish as quickly by games of chance, under the shadow of the pines, at night. The clinking of glasses, in drinking to the health and good luck of a "pard" and the dizzy whirl in the dance hall under the flaring and flickering flames of pine knots and tallow candles, were vivid scenes of the stirring times in the "good old days" of the early eighties.

The central axis or core of the Black Range of mountains extends approximately north and south and is composed of granites, gneisses and quartzites, flanked with the massive limestones and shales of the Carboniferous period.

It is probable that some of the lower series of rocks here, as at Lake Valley and Sierra Blanca, may belong to the Silurian and Devonian systems. The sedimentaries dip away from the axis of the range on either side at an angle of about 25°. Other disturbances occurred subsequent to the primary orogenic upthrust of the granite core, dislocating, warping, cutting and partly covering the great sedimentary series with dikes and eruptive sheets of porphyries. The ore deposits of Kingston in the Black Range, lie at or near the contact of the thick Carboniferous limestones and a blue-black shale. Some of these limestones are fossiliferous to a certain extent, indicating their geological horizon; the shales are several hundred feet in thickness. These stratifications dip to the east conformably with the basal quartzite. Still farther to the east beyond the mineralized zone appear sandstones and shales of a later period, which in turn disappear under still more recent rocks of an eruptive character.

Usually the ore deposits occur in the cavernous receptacles previously formed in the limestone and rarely in the true contact itself.

During a recent visit by the writer to the Illinois mine, which is a typical representation of the camp, examination bore out the fact that wherever any connection leading from a limestone cavern with the shale contact, even though it be as small as the blade of a knife, mineralization occurred; on the contrary where no such communication with the contact was found, such isolated limestone chambers were destitute of ore values. This rule may be accepted as general throughout the Kingston camp; in the Bromide (Tierra Blanca) district to the southeast, and in the Palomas district to the north. Had the mine operators observed this fact and followed the contact to where a communication into the limestone seemed to reach and then to have drifted on this opening, much of the uncertainty of finding an ore chamber would have been eliminated. On the contrary, enormous amounts of money were expended in driving long tunnels and drifts at random in the solid lime rock, without any results, excepting some discoveries made by sheer chance.

Concerning the theory of ore deposits in this camp and in other parts of the Black Range, there seems to be only one tenable exegesis, and that is deposition by descending waters. Since the limestone forms the foot wall, all descending waters would naturally reach it as it penetrated the shales at and near the contact; the openings in the limestone leading to the cavernous receptacles, would thus be in a favorable position to receive the mineralized solutions. Much of the gangue matter of the ore is evidently derived from the overlying shales, and has been deposited in the descent of the water, due to gravity. The largest and richest ore bodies have been found near the surface. A recapitulation of these observations would tend to show that deposition took place from above, by the action of circulating descending waters, which became mineralized in their passage through the overlying shales. It, therefore, appears more than probable that the source of the silver ores of the Kingston camp, and many other places in the Black Range, is from the black shales of the Carboniferous rock system.

Kingston camp holds the record of New Mexico in the production of silver. From the time of its discovery to January 1, 1904, the estimated production is $6,250,000, nearly all of which is silver. At the present time but little is being done; the production having gradually declined, since the demonitization of silver, until it is practically nothing at this writing.

The famous producers of the camp, with a number of the more prominent prospects, are here given:

Lady Franklin, Brush Heap, Illinois, Saratoga, Templar, Virginius, Calamity Jane, Superior, Monaska group, Black-eyed Susan, Andy Johnson, Old Savage, Keystone, Comet, Black Colt, Bonanza and Teddy group; the latter locations were made recently.

Palomas (Hermosa) District.

This camp lies due north of the camp of Kingston, about twenty-five miles, and took its name from the Rio Palomas, which heads in that vicinity. The general contour of the country is pretty much the same as that about Kingston, and the geological structure is similarly arranged.

All of the principal mines are distributed along contact veins; the ores carry more lead than those in the Kingston camp. Copper values also, seem to increase on going north from Kingston to Grafton; the former camp is practically destitute of copper ores. This district about Hermosa, had produced about $1,250,000, principally in silver, up to January 1, 1904. Very little activity is manifested in the Palomas district at the present time, May, 1904.

The principal work done during 1903, was on the Palomas Chief, which in the early days was a famous silver producer; this late development was disappointing to the lessees. This mine alone, is said to have had a production of nearly half a million dollars in silver to its credit.

Next to the Palomas Chief the Pelican group is perhaps the best known in the district, and has a record of production.

Among other properties which are favorably and well known in the district are the Philadelphia, Antelope, Emberlight, the "L," Atlantic Cable, American Flag, Flagstaff and Ocean Wave.

Apache District No. 1.

In going from Hermosa camp due north twelve miles, the Apache mining district is found which embraces the mining camp of Chloride.

Harry Pye, a "mule skinner," in the employ of the United States, in 1879, is conceded to be the discoverer of the district. In transporting military supplies through that section of the country, Pye found a piece of float in the canyon near the present site of Chloride postoffice, which he had assayed and found that it carried high values in silver. After fulfilling his government freight contract, with a small party of friends he proceeded to the point of his discovery, and located what is now known as the Pye lode. Due to the silver ore, the camp received the name of Chloride. The main body of prospectors arrived in the new district in 1880, and in 1881 the town of Fairview started a few miles to the northeast.

A number of prospectors and settlers were killed by the Apache Indians in 1881, among them was Harry Pye, the discoverer of silver in the district. It was due to the harassing Apaches that the district received its name.

Part of the district lies in Sierra county and the remaining portion in Socorro county.

Overlying the basal gneisses and quartzites, are the thick Carboniferous series of rock; these latter beds are much broken and distorted by dikes and sheets of andesite and trachyte flows, of profound magnitude.

At the summit of the range, is the continental divide, a rhyolite capping is observed, in which no ores of value have yet been found. Andesite-porphyries seem to be the prevailing type of eruptive rocks; though extensive tongues of trachytes reach out and cover certain localities.

The veins are usually very pronounced, and are contacts, contact fissures and true fissures.

Their general trend is north and south or parallel to the range, dipping to the east at an angle of about 70°. Silver, copper and gold ores are the prevailing types. Gold values seem to predominate in the north end of the district; while to the south end, silver and copper are characteristic features of the mineralization. In every instance the gangue matter is quartz.

In the vicinity of Chloride, the veins are usually contacts between lime and porphyry; some of which attain a very great width. The values on the whole are exceedingly low.

Within sight of the noted pinnacle, known as Victorio's Outlook, is the well known Silver Monument mine, which has a record of production of $100,000 up to 1893; having lain idle up to 1903, when work was resumed in a small way. Southwest of Chloride a few miles is the Colossal mine with about $70,000 to its credit. Nearby are the U. S. Treasury group (a big body of low grade ore), the St. Cloud group, Nana, Midnight, Readjuster, Nordhausen and New Era group. This latter group produced a few cars of rather a high grade bornite, during 1902-3; work was suspended in 1904.

Near Grafton postoffice, about two miles by wagon road northwest of Chloride, is the celebrated Ivanho property, formerly owned by Col. Robert G. Ingersoll and associates. The Ashville, Emporia, Braxton and others, were formerly producers in the early days.

To the north of Grafton some three miles are the Great Republic, Kingston, Triangle and Julia. These properties are on a very bold outcrop of quartz which can be traced for several miles. In the immediate vicinity of the Great Republic may be mentioned the Chicago, Minnehaha, Crawford, Olympia, Sunrise, Gold Bug, Camden and others which are more or less developed.

Promiscuously throughout the district are a number of fair prospects such as the Copper Queen, Bullion, Alaska, Emporia, Montezuma, Golden Chance, Golden Revenue group and the Elephant group.

Limestone (Cuchillo Negro) District.

The Cuchillo Negro range of low mountains or hills lie on the east slope of the Black Range, between and a little east of a line drawn connecting Hermosa with Chloride. On approaching the Black Range from the east, the Cuchillos present a distinct reddish appearance standing out in bold relief from the Black Range which forms the background. The Cuchillo mountains were elevated at a later period than that of the Black Range.

The Limestone or Cuchillo mining district was established immediately after the first year of excitement at Chloride; Edward's camp sprung into existence then and many claims were staked out during 1882.

Noted among the older locations are the Black Knife, German and Rifle shot. The Dictator, Vindicator, Confidence and Enterprise groups are mainly prospects and were located later.

It is in this range that a large deposit of iron ore exists, which belongs to Thos. Scales, and is known as the Iron Mountain group. Copper, lead, silver and zinc are the principal metals beside the iron.

Sullivan's Hole District.

This district or locality marks the terminus of the Black Range on the north. Here the low hills imperceptibly vanish into the great plain which stretches to the north.

Large bodies of low grade ore exist in this isolated section, which have never been critically examined by anyone, owing to the great distance from transportation.

CHAPTER XV.

MINING DISTRICT OF THE SIERRA OSCURA AND SAN ANDREAS RANGES.

Since these two mountain ranges are more or less mineralized throughout their whole extent and the occurrence of their ores appear closely related genetically and in their physical constitution, it would seem proper to group and discuss the various mining districts and camps under one general heading.

The Sierra Oscura and San Andreas ranges are very similar in appearance to the Sandia mountains, which lie immediately east of Albuquerque. They are vast tilted blocks and bear in their facial expressions, along their bold escarpments, evidence of a most profound fault; the throw being nearly one mile perpendicularly. This fault line seems to have passed through the little Burro Gap which marks the dividing point between the two ranges; for in the Sierra Oscura range lying to the north its scarp faces the plains of the Jornada del Muerto, while to the south, the San Andreas present a similar appearance fronting the east, with its precipitous face looking down on the mysterious plains of the "white sands." The ores found in these ranges of mountains are iron, copper, gold, silver, lead, zinc, mica and fluorite. All of the mining districts lying in the Oscuras are frequently alluded to as the Oscura district.

For convenience we will begin at the north end of the Sierra Oscura range and proceed south and to the San Andreas range taking the various districts and camps in their respective order.

Jones District.

This district lies in the north end of the Sierra Oscura in Socorro county and was established in 1902. P. C. Bell, an old time prospector and Fred Schmidt, who is a rancher on the Jornada plains southeast of Hansonburg, were the principal founders. The name was given in honor of F. A. Jones, the writer of this volume, who first recognized the importance of its vast iron deposits, during a visit there June 26, 1902, and who became interested with those two gentlemen in a number of iron locations during the summer and fall of that year. The iron deposits of the Jones district seem to be of more economical importance than any of the other minerals.

By referring to the chapter on iron, a brief description of these deposits is given.

About one mile southwest of the Jones iron group considerable development work was done on a copper prospect, by an Italian, with no definite results; the ore is principally a carbonate, in a small fault fissure in the upper Carboniferous limestone.

In the southwest part of the district, just below Bruton's ranch at the mouth of the canyon, H. J. Rehder and others, from Albuquerque, in 1902, did considerable development work in the Permo-Carboniferous rocks for copper, but apparently without success. Along the escarpment south of the mouth of Bruton's canyon, are several galena prospects in the carboniferous limestone; considerable zinc blende is associated with the lead. Still further south and in the vicinity of Monument peak are several good veins of fluor spar, with a sprinkle of galena running through the vein stuff.

In the inmediate vicinity and to the north of Bruton's canyon, an "old timer" by the name of John Smythe has located numerous copper prospects all of which lie in the gray and chocolate ʻsandstones, presumably Triassic. It is the opinion of the writer that nothing permanent will ever result from these latter locations.

Hansonburg District.

It was about the year 1872 when this section of Socorro county first attracted the attention of prospectors.

A man by the name of Pat Higgins, who previously sold his mining interests in Water canyon in the Magdalena mountains, appears to have been the discoverer of the district. Subsequently the district received its name from the old prospector Hanson, who formerly operated at Magdalena and Socorro.

Prospecting has been carried on intermittently ever since the district became known; thus far but little success has attended the efforts of the prospectors.

The Alcazar Copper Company during 1901, under the supervision of A. J. Borden developed one or two properties quite extensively, having erected a gasoline hoist and numerous buildings; but during the latter part of the year trouble arose in the management among the eastern stockholders and the concern suspended work indefinitely. One car load of ore, only, was shipped by this company.

The Hansonburg camp lies on the west side of the Sierra Oscura in Socorro county, near the center of the range, in the series of low red hills which lie along the base of the mountains; these red hills extend as far north as the red tanks west of Bursum's sheep ranch. This sequence of rocks is copper bearing and belongs to the Permo-Carboniferous series.

Estey City District.

Across from Hansonburg over on the east side of the Oscura range is the Estey City district. This district sprung into existence in 1900; although copper had been known to exist there many years before.

During a visit to that region by the writer, in June 1902, the time for a cursory examination was taken and a section was hastily sketched, which presents a general idea of the peculiarity of the copper deposits on this side of the range.

The red hills or series of rocks on this side are similar to those on the Hansonburg side. There are three distinct mineralized zones, two in sandstone and the third in a gray shale; all of which carry copper compounds. Copper glance (chalcocite), copper carbonate (malachite), copper pyrites (chalcopyrite) and peacock copper (bornite) are found; the latter ore occurring in nodules as large as two inches in diameter. In the shales of the Lucky Jack location such nodules are abundant.

Since these red series of rocks immediately overlie the upper Carboniferous, they no doubt belong to the Permian.

The cupriferous solutions seem to have precipitated and segregated about organic substances, since the impressions of fossil leaves, stems of plants and parts of trees are found; the copper compounds having replaced the carbonaceous material, and thus preserved the texture of the plants. Owing to the occurrence of these regular beds or zones of copper ore, it would seem probable that the copper was contained in the water that carried the enclosing sediments. Further to the east, about fifteen miles, in the vicinity of Oscura station, on the El Paso and Northeastern Railway a fourteen inch vein of coal was being developed by the Estey Mining and Milling Company, at the time of the writer's visit.

On the whole, the ores of the district are very low grade, not exceeding four per cent in copper. The area over which the copper extends is considerable, aggregating several thousand acres.

The Estey Mining and Milling Company was promoted by David M. Estey of Owosso, Michigan, who died in September, 1903, in New Haven. At the outset this company practically covered the whole country and at one time controlled, it is said, about three hundred claims.

A large electrolytic plant of 100 tons capacity was installed, water piped from Moonshine three miles distant, a townsite laid out, numerous buildings erected and other magnificent expenditures indulged in; all of this was done before the ore bodies had been properly exploited or a method for the correct and economical treatment of the ores had been investigated.

About $200,000 were spent in this manner before the crash came which caused the company to suspend operations.

The Dividend Mining and Milling Company, a new organization has recently been effected and is now rehabilitating the wreckage and mistakes of its predecessor. The northern part of the district is principally owned by the Sierra Oscura Company, while the southern portion belongs to the Dividend Mining and Milling Company, successor of the Estey Mining and Milling Company. The Little Effie, owned by D. Doherty and the Just-Before, owned by Capt. Roberts, deserve mention.

The production of the district up to January 1, 1904, is approximately $10,000; the values being chiefly in copper.

If a success is made working these low grade ores, it will

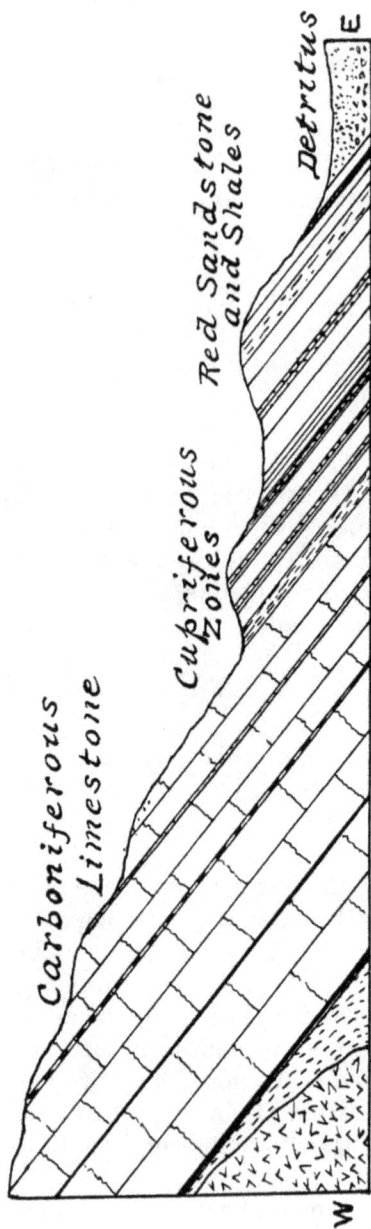

Fig. 18 - SECTION ACROSS THE SIERRA OSCURA.

require the greatest metallurgical skill and the strictest economy.

San Andreas District.

Under the San Andreas district, are a number of smaller sub-districts which refer to particular localities in the mountain range.

In the Little Burro Gap are some lead and copper prospects; one copper property has been quite extensively developed and is patented.

The Chicago, a lead claim, was located in the fall of 1903 by Fred Schmidt, F. C. Toney and the writer, which shows fairly well for a prospect; it is in the upper carboniferous limestone and near the contact of the Permian.

To the south of the main wagon road in going through the gap and on the west side of the divide there appears to be favorable prospects in zinc-lead ores. In an old working about twenty feet deep, sphalerite seems to predominate over the galenite.

Across the divide and in the vicinity of Mocking-Bird Springs and Mine, much prospecting has been done in the early days. Lead, zinc and copper sulphides are the prevailing mineral characteristics.

The principal property here is the Mocking-bird group, of twenty-three claims owned by the Dividend Mining and Milling Company of Estey City.

Around Capitol Peak, still farther to the south, copper seems to become more prominent. In the Mocking-Bird district as a general rule, the vein-stuff carries considerable fluor spar. Immediately north of Mocking-Bird springs in the first canyon there are indications of mica, which is alluded to under the caption of mica.

West and a little south of Capitol Peak, in the vicinity of Dripping Springs, copper and lead predominate. Several properties in this region have been extensively developed, though none have, as yet developed into producing mines of any consequence.

The Boulder group and Whirlwind, Coyote and Rattler, are managed by Thos. T. Leask; the Bean group is owned by S. L. Bean, and the Grand View group, 16 to 1 group and Gold

Reserve deserve mention at this time, and seem to have a future.

Mound Spring District.

This district is situated in township 6 south, range 6 east, and lies midway between Estey City and the Jones district, on the eastern slope of the range. It derived its name from some springs in that region.

Nothing has been done here for a number of years; two patented claims include everything worth mentioning. The ores are gold, silver, copper and iron.

CHAPTER XVI.

SOCORRO MOUNTAIN DISTRICT.

This district which lies immediately west of the town of Socorro, was a scene of much mining activity during the eighties. The Rio Grande smelter situated some two miles west of the court house of Socorro, was in full blast then; receiving its principal fluxing ores from the famous Kelley and Graphic mines at Magdalena, and the more silicious products from Socorro mountain. This once prominent smelting plant is now practically dismantled; and with the dying fires of its stacks the life of the Socorro district passed out.

At present there is little prospect of a revival of the district, owing to the highly silicious character of the ore and the low price of silver, inactivities will likely be prolonged for some time to come.

The geology of this district is rather complex; the origin of the mountain itself is due to an old andesite volcano, from or near which several distinct flows have emanated. Taken in their order relative to age, the first outbreak was evidently andesite, which is practically covered by subsequent flows; and can be seen in only a few places where erosion has stripped or cut through the covering. Then comes the trachyte, after which the rhyolite and then lastly the basalt *mal pais*. Periods of erosion of variable duration occurred between the flows.

Below the precipitous cliff which fronts the Rio Grande and to the north of the principal mines of the district may be seen a superficial exposure of the lower Carboniferous limestones; the same series may be seen about 15 miles farther north at the Lemitar mountains, in a similar exposure.

On top of the highest peak of the Socorro mountain some typical phonolites exist: the general type of rock on the peak however, is trachyte. By an increase in soda the trachyte passes into phonolite; this point was the only place in the district in which the latter rock was observed. South of the principal mines and to the north of Blue canyon are extensive beds of fire clay. This clay seems to be the result of the kaolinization of the old andesite flow. It is on the line of this contact between the andesite and trachyte where the

Fig. 19—SOCORRO MOUNTAIN. Photographed by F. A. Jones, 1902.

principal zone of mineralization of the district lies. The ore is a white and gray quartz, carrying silver, but no gold of which to speak.

The mineralized zone extends parallel with the base of the mountain and pitches into it at an average angle of about 45°.

It seems that some evidence of ancient workings, presumably by the Spaniards, existed in this region at the time the country was first being prospected, which was as early as 1867.

A man by the name of Hanson, one of the former owners of the Kelly mine at Magdalena, discovered the Torrance mine; he being among the first prospectors who created the excitement over the discovery of silver in the Socorro mountain. The Merritt mine was next to the Torrance in importance. To these two properties the chief production of the district is due.

The Hammel mine, lying just north of Blue canyon, carries silver and lead values; it is claimed that a few assays running high in gold were taken from this property. The Volcanic Mountain group, consisting of nine claims and located higher on the mountain, was thought to have considerable merit by the former owner, William Glasson, now deceased.

Most of the properties in Socorro mountain are now either abandoned or merely kept alive by the annual assessment; no effort is made toward mining, whatever. The production of the district since its organization to the present time, January 1, 1904, is $760,000.

Socorro mountain is clothed in grandeur to the student of nature. The several flows of brecciated and complex eruptives bear testimony of the awful ordeal through which the region has passed. Considering the recent earthquake shocks in this immediate section in connection with the thermal springs at the base of the mountain, these point to the fact that life is not yet extinct in that shattered and rock riven member.

Lemitar, Mountain District.

This locality lies to the north of Socorro mountain about ten miles. The country is broken, faulted and titled, bearing evidence of much aqueous and igneous activity, as though the first volcanic outbreak came through a marginal sea bottom. Andesite, trachyte and rhyolite flows followed in pretty much the same order as at the Socorro mountain. At or near the contact of the acid members with the limestone, a mineralized zone exists which carries the sulphides of lead and zinc.

Considerable prospecting has been done in the Lemitars, yet nothing has been found that would justify a heavy expenditure in developing the existing prospects. Some very good assays in silver have been reported from certain localities in the district. On the northern slopes the tendency of the mineralization is toward copper.

San Lorenzo District.

This is comparatively a new district; although the presence of copper, gold, silver and manganese have long been known to exist. The district took its name from the San Lorenzo pass, which is immediately south of the Jerome copper property.

It lies about eight miles west of San Acacia, a village on the A. T. & S. F. Railway and about twenty-three miles northwest of the town of Socorro, in Socorro county and on the Servilleta (La Joya) grant. All matters of business pertaining to the grant are transacted through a Board of Supervisors elected by the resident owners of the grant.

The Board of Supervisors is usually very fair in dealing with *bona fide* prospectors and mining men, desiring concessions and rights.

The most important discovery in the district—the Jerome lode—was made in the spring of 1901, by Mark Thomas of Albuquerque.

A number of locations have been made at various times in different parts of the district; but, the most promising of them all is the Jerome group of copper claims. The Jerome, the Camp Bird and the Sacramento constitute the best locations so far as now known.

This property is yet only meagerly developed. The work on the Jerome claims consists of two tunnels or drifts about 35 feet apart and some 40 feet each in length. A cross-cut of about 25 feet is run from one of these to intercept the drift which is connected by a 40-foot shaft.

Several shafts of superficial depth are on the Sacramento; one is down about 25 feet. A few open cuts constitute the work on the Camp Bird.

Native copper, cuprite, chrysocolla, malachite, tenorite, with small values of gold and silver, embrace the principal minerals and metals. Some compound minerals of uranium and vanadium ores seem to occur sparingly, which would suggest the idea of the presence of elements having radio-activity.

Owing to the peculiar manner in which the native copper occurs, embedded in lime-spar, in shot-like and acicular aggregations, it would indicate secondary deposition from the sulphides, oxides and possibly carbonates which have been leached from the encasing amygdaloidal rocks.

The theory advanced by the workmen and owners, that the copper had been melted by volcanic action and forced upward from below, is not well founded.

This district contains much manganese and may become famous for that mineral when properly exploited. To the north, northwest and west of the Jerome group a favored field for manganese ores undoubtedly exists; as well as in the immediate vicinity of the Jerome property. Pyrolusite seems to be the prevailing type of manganese ores. The geology of the district is comparatively simple in structure, though interesting.

The sedimentary beds are tilted and dip to the west at an angle of about 25 degrees, the same as in the Lemitar mountain, a few miles to the south.

A reddish amygdaloidal volcanic rock encases the mineral-ized vein of the Jerome, which seems to be in a fractured zone along a fault fissure. A system of faults extends north and south and parallels the main ridge. Horizontal faulting, also, exists and mineralization occurs along these planes, as well as in the main fault-system.

More than a mile to the east a similar eruptive is found which conforms to the sedimentaries and in which the occurrence of copper is noted.

Four or five miles north of the Jerome an extensive flow of *mal pais* is impregnated with copper in the form of carbonate, which will average about three per cent of copper. It appears that the solution of copper circulated through the lava just after or about the time of the final cooling of the flow.

Hanson District.

This district embraces the Sierra Ladrones, and lies thirty-five miles northwest of the town of Socorro.

It was discovered by a man by the name of Hanson in 1868, one of the owners of the Kelly mine in the early days.

Owing to the rugged character of the district and the scarcity of water in that region, but little prospecting has been done there since its discovery.

The principal peaks appear to be granite; while at the base, porphyries and eruptive dikes abound.

Prospectors have oftentimes reported rich gold discoveries in the Sierra Ladrones, but were never able afterward to return to the coveted spot, after once leaving it for water and supplies.   Others, who have wandered into those mysterious fastnesses, are said to have never returned.

Since the only inducements held out to those who dare enter this enchanted region, were disappointments and death, no one cared to take chances against such odds.

Since the year 1900, however, considerable prospecting has been going on and some fine looking prospects in gold, silver and lead are now being developed.

The district has never produced.

Canyoncito District.

About five miles to the north of the village of San Acacia in Socorro county, and on the opposite side of the Rio Grande is the Canyoncito mining district.   The camp lies in the Servilleta (La Joya) grant, which is a community grant.

The country is broken and has been but little prospected, owing to its out-of-way situation.

The basal formation is granito-gneiss, carrying large crystals of biotite and hornblende.   The axial ridge is parallel to the Rio Grande.   To the northeast about five miles is a basin which belongs to the cretaceous coal measures, and which has never been exploited to any extent.

The Carboniferous limestones and shales lie next to the gneissoid base and dip at quite an angle toward the river; this in turn is flanked by the Permo-Carboniferous.   Next to, and even covering this latter series, occur thick beds of gravel and conglomerates, bordering the flood plain of the Rio Grande.

The minerals occurring in this district are lead, iron, manganese and also, some copper, gold and silver. Only one group of mines has been worked to any extent—the Dewey group—which consists of three claims.

This property is a low grade galena ore, carrying but little values in gold and silver. The gangue matter is calc-spar and quartz; the lime giving way to the quartz at depth. The vein is a remarkably smooth fissure, the sides of which are almost mathematically true; the strike is approximately east and west and the dip is 80° toward the south. The width of the vein is about an average of four and one-half feet; the character of the wall rock is the same on each side.

A shaft is down to a depth of about 110 feet; which follows the vein the whole distance. Some little drifting has been done, but no stoping. This vein has a strong outcrop for a distance of over 3,000 feet. The Dewey lead property was discovered by Lon Jenkins in 1895. The iron in the district has been prospected some; though mainly for the gold it contained.

The copper has never been investigated carefully, notwithstanding there has been some very rich pieces of sulphides found on some of the prospects.

Chupadero Mountain District.

These mountains lie to the east of the Rio Grande in the north-east part of Socorro county. But little prospecting has ever been done in them; they are low and seem to be composed principally of Cretaceous and the red series of sandstone, broken at intervals by eruptive dikes.

Indications of copper are found in various places; in one instance, in particular, the sandstone is impregnated with malachite and the cupriferous sandstone resembles that found in the vicinity of Las Vegas. Lack of water and great distance from transportation render this section a comparatively unknown quantity.

Rosedale District.

Rosedale camp lies at the north end of the San Mateo

mountains, in Socorro county, about thirty miles west and a little north of San Marcial.

The first white prospector in the San Mateo mountains and to whom the discovery of gold is due, was J. W. (Jack) Richardson, who still resides there, in Rosedale. This was on December 15, 1882, Mr. Richardson coming into the San Mateo mountains by way of San Marcial. A little later a rush was precipitated in that direction by prospectors, which was shortly afterwards checked by the frequent incursions of the Apache Indians under the leadership of Geronimo. Several massacres occurred and most of the prospectors quit the country.

Comparatively little prospecting has been done in the Rosedale district since the first excitement, until quite recently.

The principal mine that made the camp famous, is the Rosedale. This was the first discovery and mining claim located in the district; it was made by Jack Richardson and wife. Mrs. Richardson, it is claimed, found the first float, which she prevailed on her husband to have assayed, and which proved to carry good values in gold. The Rosedale mine at the present time belongs to the W. H. Martin Company, this management having operated it for several years. This is one of the few properties that has done development and paid dividends, without stoping or taking out ore, other than that encountered in sinking and drifting. A depth of 750 feet has been attained with levels at every 100-foot interval.

The vein is a fissure encased in rhyolite, this type of vein and enclosing rock appears to be general throughout the district. The ore is free milling near the surface, but is practically base in the deepest working.

A 10-stamp mill with a cyanide plant constitutes the metallurgical equipment, which was completed in 1898.

The Rosedale mine is among the leading gold lode producers of New Mexico.

Other properties in this district are receiving attention and are being systematically developed in their working and in the installation of machinery; the New Golden Bell and the White Cap deserve especial mention in this respect.

Other well known and favorable prospects are here given: Bay Horse, Ninety-nine, May Dew, Baking Powder, New Year, New Year No. 1, Rockefeller, Golden Gate, Gold Cap, Amy B., Greenwood group and the Graham lode.

Red Hill District.

Due west of Rosedale about fifteen miles, in township 6 south, range 8 west, is the Red Hill district.

This is an unimportant camp, very little if any work has been done there since in the eighties of the past century.

Gold and copper are the principal values contained in the ores.

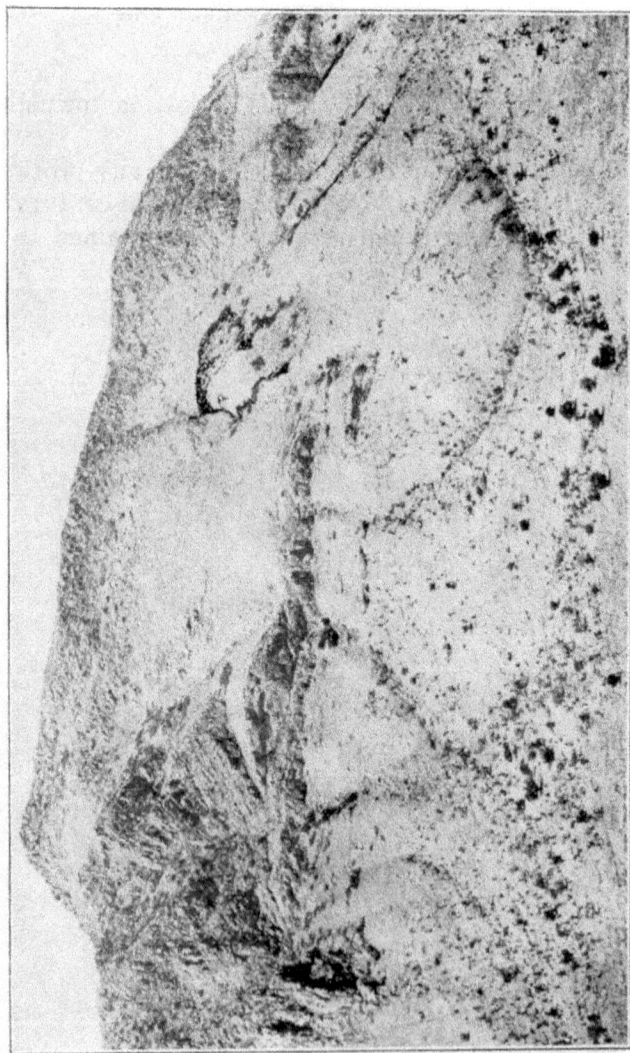

Fig. 20 MOUNT MAGDALENA, SHOWING THE FACE OF MADELINE. Photographed by W. M. Borrowdale in 1904.

CHAPTER XVII.

MAGDALENA DISTRICT.

This region stands pre-eminent, in New Mexico, in the production of lead and zinc.

The Magdalena district lies west from the town of Socorro, at the end of the Magdalena branch of the Atchison, Topeka & Santa Fe Railway, in north central Socorro county. It took its name from Mount Magdalena, which bears a face or image formed by a growth of shrubs circumscribing a small area of rock debris on the east side of the mountain slope. This remarkable growth of shrubbery forms with distinctness the outlines of a woman's head.

The early Spanish explorers pictured in this benign face, the image of Our Lady Magdalena.

A beautiful legend is handed down to us that no murder has ever been, or could be committed under the compassionate gaze of that contrite countenance. This spot became a place of refuge and the savage Indian would forego his bloody deeds under the shadow of the holy mount. Those in danger would flee to this enchanted spot, and thus, became invulnerable from the onslaught of their pursuers.

The discovery of the Magdalena district dates among the earlier discoveries in the latter half of the nineteenth century.

Col. J. S. Hutchason, better known as "Old Hutch" among the mining fraternity, is conceded to be the pioneer of the district and the one who made the first location in the camp. He and his peon, a Mexican by the name of Barado Fidey, were in that region looking for rich float which Pete Kinsinger* had found at Pueblo Springs during the war. Failing to find the object of their search they turned their attention toward the Magdalena mountains where they found rich lead croppings.

The day following the Juanita lode was staked out, which

---

*Kinsinger was a pioneer miner at Elizabethtown the time gold was discovered there.

was the first claim located in the district; this was in the
spring of 1866.   About three weeks later Hutchason located
the celebrated Graphic mine.

‥ Active mining began a little later, the lead ore was smelted

Fig. 21—"OLD HUTCH'" DISCOVERER OF MAGDALENA DISTRICT.
Photographed by W. M. Borrowdale, 1904.

in an adobe furnace and the bullion was hauled to Kansas
City by bull teams.   Hutchason found the Kelly mine and
turned it over to his friend Andy Kelly to locate, who was
operating a saw mill, in now what is known as South Camp.

The Juanita was subsequently sold to Col. E. W. Eaton and

associates; the Graphic mine was sold for $30,000 later on.

This latter property passed through several hands and was recently sold, in the latter part of March, 1904, to the Graphic Lead and Zinc Company, which also, operates mines in the zinc regions of southwest Missouri.

Hutchason afterward jumped the Kelly property, the owners failing to do the necessary annual assessment; afterward he sold the property to Hanson and Dawsey. These parties in turn relinquished their holdings to Mr. Gustav Billing for $45,000. Billing after assuming control of the property erected a smelting plant at Socorro where the product from the mine was treated. The Magdalena branch of railroad was built under the guarantee that the Kelly mine would furnish a certain tonnage for a definite length of time. The production of the Kelly and Graphic mines up to January 1, 1904, is approximately $5,800,000; with the Kelly in the lead. This production was entirely from a lead-silver ore. During 1904, new life dawned on the old lead camp; large bodies of smithsonite and other forms of zinc ore were exploited, which hitherto had been overlooked. It was on the strength of the zinc ore bodies which effected the recent sale of the Graphic mine to the zinc operators in Missouri.

The exploitation and uncovering of zinc gives the Magdalena district a new lease of life, which promises to be as permanent as that of the lead. It appears that the carbonate lead ore is practically exhausted, insofar as the Kelly and Graphic mines are concerned. In the Juanita, Ambrosia, Young America, Enterprise, Juanita South and several other properties, large and profitable bodies of zinc are being uncovered. Formerly these rich zinc carbonates were thrown over the dump as being worthless; many of these dumps are now being overhauled with remunerative results.

The Hardscrabble mine was located by a man by the name of Baker, who sold it in the year 1868 for the sum of $25.00. The peculiarity of the geological position of the limestone of the Hardscrabble, which is so intimately associated with the ore of the district, is interesting in the extreme. This limestone is well nigh the top of the mountain and apparently embedded in the solid granite and dips toward the east, which is opposite from the dip at other points on the west side of the range. It is probable that in the general upheaval of the range this patch of limestone became detached, and encased in the plastic magmatic mass, finally settling to its present position.

Since the lime terminates abruptly in the granite, the existence of lead carbonates also ceases. In the granite fissures, however, good values in sulphide of copper were found. The Hardscrabble is among the most noted properties in the district; it is credited with a production of $325,000 up to January 1, 1904.

The country to the west has been the scene of much eruptive activity; in fact the Magdalena range was the central theatre of volcanic phenomena which extended to the adjacent surrounding country.

Two principal craters, Big and Little Baldy, bear evidence of the baptism in fire during that period.

Successive flows of andesites, trachytes, rhyolites and basalts are observed in various parts of the region; the later flows, in many instances have entirely covered the older lavas. The massive Carboniferous limestones, which flank the west side of the range, are fractured and dislocated in such a manner as to form a general system of fault fissures; in these faults and along the lime contact exist the principal ore bodies. The commercial ores common to the district are mainly lead and zinc carbonates and sulphides, with some silver and compounds of copper.

It is observed that in every case where good bodies of ore are found, that the limestone has been influenced by eruptive dikes and sheets of a acidic character.

Beside the claims already mentioned the district abounds in smaller producers and favorable prospects, which may be enumerated in the following properties:

The Cavern, Fifty Eight, Tip Top, Grey Hound, Samson, Review, Ouray, Umpagra, Legal Tender, Silver Peg, Silent Friend, Cimarron, The Nit, Imperial, Grand Ledge; Pearl, Key group, Oxide group, Wheel of Fortune, Golden group and Iron Mask. This latter property is situated in South Camp and embraces in its connection the Old Iron Mask Smelter, which was erected but was never "blown in." The Graphic Smelter, between the villages of Magdalena and

Kelly and which was operated steadily for a number of years, has been practically idle since 1900. The Magdalena district has produced in values all told, up to January 1, 1904, $8,700,000, in lead and silver.

Pueblo District.

The discovery of rich silver float in this district by Pete Kinsinger, a soldier, when stationed at Pueblo Spring in 1863, led to the discovery of the Magdalena mining district a few years later. The Pueblo district lies a short distance immediatly north of the town of Magdalena. Many locations were made and much prospecting done here in the latter sixties of the past century; nothing of any importance was ever found, with the exception of some small rich veins of silver. The formation is eruptive, breaking through the underlying limestone, and composed chiefly of trachytes and rhyolites, traversed by numerous dikes of a dark green color.

The veins are very irregular, shattered and faulted and do not appear to reach any great depth; the Red Jacket being an exception with regard to depth. The Chloride lode, a patented property, is a contact fissure lying between lime and porphyry. It is only meagerly developed, having a shaft sixty feet deep. In the Sophia location much of the silver was native, and existed in needle or acicular shape, which was rather a remarkable occurrence.

The celebrated Ace of Spades mine was located by George Brown in 1868, who previously was one of the California troopers of the Civil War, under General Carleton.

Quite a sum of money was spent on this property and a number of soldiers of the United States Army became interested during the period of its early development. Its production is not known.

About three or four hundred yards northwest of the Ace of Spades is the outline of an old prospect pit, supposed by some to be of Spanish origin.

Iron Mountain District.

Uncle Billy Hill, as he was familiarly known in the early days of mining in New Mexico, was the discoverer of silver ore in the Iron Mountain district. He came into the Iron

mountain region in January, 1881, and made the first location
in the camp, known as the Summit lode.

A little later in the season Davis and White drifted into
that region and located the Old Boss.

In June, Judge Hagan of Socorro became interested with
Uncle Billy in the Summit claim; after which a considerable
quantity of hard carbonate lead-silver ore was mined, for
which the district is noted.   A little of the best grade of
the Summit ore was sent to the smelter with fairly good
results.   The occurrence of the ore is remarkable, inasmuch
it is found in a pipe, and is termed a "pipe vein," which has
been followed down to a depth of over 300 feet without reaching the end.

During September, 1881, G. L. Brooks and F. J. Wilson
came into the district and bonded from Davis and White the
Old Boss for $17,500, which at that time was a shallow pros-
pect hole, and constituted the principal development.

These two gentlemen, by February, 1882, paid off the
amount of the bond and sold a half interest in the property
for $25,000.

The ore of the Old Boss is a hard carbonate of lead and
silver and averaged $250 per ton, the quality shipped at that
time; a number of carloads of a lower grade of ore is now on
the dumps.

About 800 feet of work has been done on this property
and the greatest depth attained is 165 feet; the mine has lain
idle a number of years.

In the fall of 1881 C. T. Brown located the Mammoth group
of claims, which he developed and subsequently sold.   The
Cabinet mine is also well known in that region and is extensively developed.   The ore is said to occur in a pipe vein
here, very similar to that in the Summit.

An iron lode has also been prospected in the district to a
depth of eighty feet.

The district is in Socorro county, about ten miles west and
a little north of Magdalena, and eight miles due west of the
Pueblo district; it borders on the San Augustine plain.

The name of the district was suggested by the immense
quantity of iron float found scattered over eight square miles
of that region.   The iron is found in pieces from the size of a

pea to chunks of several hundred pounds. Every piece of the iron is smooth and well rounded, indicating that is has been transported by glacial or river action from a distance.

Most of this float was picked up a number of years ago by the Mexicans and hauled to the Rio Grande smelter at Socorro for flux; it is mainly a high grade hematite, but possesses some magnetic properties.

This region is sometimes spoken of as the "Ten Mile" district.

A townsite was staked out by Judge Hagan, Brooks, Wilson and others in October, 1881, and nearly one hundred cabins erected; lots sold as high as $250 each. The new town was called Council Rock; named from a large rock near a fine spring of water. This particular spot in the western wilds was known long before, and had become famous as a rendezvous of the Indian braves and whites, where they would sit in council to determine war or peace.

After a lapse of two decades the towering rock and the purling spring appear as of old; but, the village of Council Rock exists only in the history of pioneer mining.

Cat Mountain District.

This district lies about twelve miles southwest of Magdalena. It was first prospected in the early seventies by E. L. Smart, Pat McLaughlin and others, but without success at that time.

Later Pat McLaughlin located a number of claims which he subsequently sold to the Socorro Gold Mining Company for $25,000. This company eventually acquired a group of fourteen claims, after which it erected, in 1902, a 20-stamp amalgamating mill with a cyanide adjunct for treating the tailings.

This mill was operated only a short time in 1903, when it was closed down and is idle at the present writing, June 14, 1904. The Cat mountain proposition is low grade and mainly a refractory ore; the trouble appeared to be in not being able to save the values. A few other locations exist in the Cat mountain district, on which the annual assessment is being kept up; the Legal Tender group belongs to this class.

In the Cat mountain properties the veins are fissures and

the vein matter is principally quartz, encased in augite-andesite.

Silver Mountain (Water Canyon) District.

The discovery of mineral in this section of Socorro county is contemporaneous with that on the Magdalena side of the range.

Water canyon is the principal aqueduct of the water which falls on the northeast slope of the Magdalena mountains.

Patrick Higgins, an ex-soldier, was the first white prospector to give attention to the minerals common to this region.

He came to Water canyon in 1868 and located claims numbered 1, 2, 3, 4, and 5 which he afterward sold to a man from Colorado by the name of Justice in 1872, for $2,500.

Mining has been conducted there ever since the time of Higgins with comparatively little success. The district has been unfortunate in the promotion of mining enterprises; inasmuch as the manipulators of the mining properties seemed to be successful in saddling onto the purchaser worthless prospects and stocks which were invariably a net loss to the investor.

Injudicious expenditures in building costly plants, unadapted to the proper metallurgical treatment of certain kinds of ores, and that too before the properties were systematically and thoroughly developed, may be cited in the case of the Timber Peak promotion and management. This failure not only ruined many innocent victims financially, but has done incalculable injury to the whole district and cast a blight over the camp which will take years to overcome.

Water canyon appears to possess many encouraging prospects, which no doubt in a number of cases may prove to be producing and paying mines. The ores on this side of the Magdalenas are quite different from those on the west or Magdalena side. Here, on the east side of the range, gold, silver and copper values seem to predominate. What lead ore that is found in the district is entirely a sulphide and the association of any zinc is also of a like character.

It would appear from observations of the district made at various times by the writer, that the principal values in the ores to be sought on this side of the range and the only ones that are ever likely to pay, lie in gold, silver and copper, with the chances of success decidedly in favor of the gold and silver.

That part of the district lying well up to the west and south is more favorable to gold and silver; the lower or eastern portion runs to copper with some lead.

The Timber Peak Mining Company in the early part of the season of 1900 completed a 150 ton roll-crushing concentrating plant enclosed in a steel building. The plant was operated only a short time and then closed down indefinitely.

The plant was dismantled the following season and a greater portion of the machinery was shipped to Mexico; the building was removed across the range to Cat mountain.

Several causes are assigned to this failure; but it seems certain that the blame should not be laid to the mine. The Timber Peak ore bodies are the largest in the district, but must be classed as low grade; beneath the superficial oxidized zone heavy sulphides are encountered.

The ore is about equally divided in values between gold and silver.

The Buckeye group of mines is situated in the lower or copper zone of the district. It is a contact deposit between lime and porphyry. Considerable work has been done on the property and a depth of 300 feet has been attained. Water became a very serious factor to handle; work was suspended in 1901 and the lower workings are now filled with water.

One of the best appearing properties in the district, from surface indications and from the meager development done, is the Iron King group. It is a fissure in porphyry and carries values in gold and silver; the proposition is low grade with a chance for opening up large ore bodies.

The Little Baldy group consists of nine claims which are being developed by a tunnel, now 1,200 feet in length.

The contact between porphyry and limestone is quite favorable for a good mine; the metallic values are lead and copper.

This property belongs to the Abbey Mining Company, with Nathan Hall manager.

Another promising group which is now becoming a producer, is the Wall Street. A Huntington mill has been placed in position during the season of 1903, to treat the oxidized ores from this group by amalgamation. Several runs have been made up to the time of this writing; the results are not yet known. Copper and iron sulphides are present in certain parts of the vein. The vein is a contact between poryhyry and lime.

The Jennie Lee group has been perhaps the best shipper in the district; it lies on the east side of Copper Creek, which is recognized as the lead portion of the region. This latter group consists of four claims: the Jennie Lee, Bologna, Pat Savage and Black Veins. As usual, for this character of ore, the depesit is along the contact of lime and porphyry.

Abbey District.

In the region about Abbey the country is very much faulted and broken similar to that toward the southeast in the Pueblo district.

This district has been prospected but little, comparatively speaking; the principal development being on the property of the Abbey Mining Company, under the supervision of Nathan Hall. Copper and lead constitute the principal metallic values of the region.

A splendid quality of coking coal exists in the immediate vicinity; the extent of the field is not known. The complex nature of the faulting would render the coal expensive to mine.

The district as yet has had no production; practically nothing is being done at the present writing.

Abbey is situate about eighteen miles northwest of Magdalena in township 1 north, range 5 west, in Socorro county.

CHAPTER XVIII.

COONEY DISTRICT.

A few words concerning the history of the man whose memory is perpetuated in the name of the district, would seem legitimate and proper.

James C. Cooney, Quartermaster Sergeant of the 8th United States Cavalry, while stationed at Fort Bayard in 1875, under the command of Col. Davis, was held in high esteem, by his superior officer, due to valuable services rendered on the western frontier, as scout and guide. On one occasion, Lieutenant George W. Wheeler and his corps, then engaged on the Geographical Survey West of the 100th Meridian, were surrounded on a mountain in Arizona by two hundred Apache Indians, a courier in the party managed to escape through the hostile line and reached Fort Bayard and reported the distress of Wheeler's party. Acting immediately on this intelligence, the late Major Bob Swan, who died at Silver City, detailed twelve men, placing Cooney in command, with orders to proceed to the rescue with all possible speed.

Arriving on the scene, the Indians faced about whereupon Cooney gave the command to charge, each man selecting his quarry wherever found. A number of the Indians had guns, but the majority had bows and arrows. At this juncture Lieutenant Wheeler made a sortie with his little band, and the Indians retreated in dismay, carrying off their wounded and leaving dead seventeen of their braves on the field of battle.

During the engagement Cooney had his horse shot and received three arrows through his clothes; while three of his men were slightly wounded by arrows.

For his services on this occasion he was offered a commission, but declined on account of having discovered* during the same year 1875, high grade silver and copper ore in the

---

*A German in 1870, is said to be the original discoverer of the district.

Mogollon mountains, on a previous scouting expedition. Immediately at the expiration of his services as scout, he organized a party to prospect the district of his new discovery. The party was composed of Geo. W. Williams, Frank Vingoe and John Lambert of Georgetown; Harry McAllister, William Burns and George Doyle of Central.

Several fights with the Apaches occurred while they were engaged in making their locations; this was in the spring of 1876. On account of the increased hostilities of the Indians, no assessment work could be performed at that time; all locations were necessarily abandoned, excepting the Albatross which was located by Burns, and who managed to do his annual assessment.

In 1878 all of the original claims were re-located, excepting the Albatross, as heretofore mentioned.

The famous Silver Bar, or as it is better known, the "Cooney" mine, was among the re-locations in 1876. From the time of the second location until April 1880, work was prosecuted with considerable vigor.

It was on the 28th day of April, 1880, when the settlers on Mineral creek, now the mining camp of Cooney, were attacked by the Apache Chieftain, Victorio, and his savage band; two miners were killed. One by the name of Buhlman was found dead at the mouth of his prospect tunnel; the bones of the other person were discovered two years later at Silver Peak. Two miners, George Williams and James Taylor escaped, the latter being severely wounded; they fled to notify the ranchers in the valley of the Frisco, about seven miles away.

Roberts ranch on the west side of the stream was the rendezvous of the surprised settlers and miners. This second attack was made on the 29th day of April, and raged all day with fury; one miner by the name of William Wilcox was killed. The house was a crude stockade affair made by setting one end of small poles in the ground; yet, it served as a great protection from the missiles of death dealt out by the redskins.

A number of Indians were killed by the imprisoned whites, who fired their weapons through the openings between the poles of the cabin with considerable execution.

On the following morning, April 30th, no Indians were in

sight. Al Potter and John Motsinger decided to go up to Mineral creek and ascertain the fate of those who had no chance of reaching Roberts ranch. They had not proceeded more than half way, when they were fired on from ambush the rifle which Potter held was shot from his hand. Being

Fig. 22—TOMB OF J. C. COONEY. Photographed by F. A. Jones, 1904.

men of nerve they pulled their pistols and charged through the line of Indians, who were on foot, making their way back to the ranch.

Later in the day Cooney and William Chick tried to reach the mining camp, when both were killed.* Their bodies

*The inscription on Cooney's tomb bears the date April 29th, for the day on which he was killed; according to Mr Meeder who keeps the Dry Creek stage station, and who was one of the besieged persons at Roberts ranch is positive that April 30th is the correcte date.

were horribly mutilated.   Near the spot where Cooney was killed, a large boulder was tunneled into and the fragments of his body were laid to repose in this rock-ribbed sepulcher.

Cooney was forty years old at the time of his demise. Capt. M. Cooney, a brother of the unfortunate scout and frontiers man, arrived in the district, from New Orleans, shortly afterwards and began active development of his brother's property—the Silver Bar or Cooney mine. The Silver Bar was sold by Capt. M. Cooney to the Silver Bar Copper Mining Company, composed of Colorado parties, in 1898, who worked it at a great profit until 1901, when the property was bought by the Mogollon Gold and Copper Company. This latter company, the present owner, has done extensive development. A main shaft is now down over 600 feet with levels every one hundred feet; an air compressor and a one-hundred, ton crushing and concentrating plant are among the improvements made by this company.

The production of the Cooney mine to January 1, 1904, is approximately $1,000,000.

The Mogollon Gold and Copper Company owns five groups of properties: First the Florida group, comprising the Florida Christmas, Laura and Mame mines; second, the Independence group, consisting of the Independence, Postmistress, Bloomer Girl and Nancy Hanks, Jr.; third, the Little Johnie group, embracing the Little Johnie, Ninety Eight and Hidden Treasure; fourth, the Malachite lode; and fifth, the Little Charlie group, containing the Little Charlie, Combination, Little "Jiant," Lena, Sandy, Homestake, Little Katie, Iron Hat and Selma.

Under the able management of W. J. Weatherby, the superintendent, the Mogollon Gold and Copper Company is developing its holdings in a scientific and workmanlike manner.

Notwithstanding the eighty-five miles haul by wagon to Silver City, the nearest railway point, many of the Mogollon mines have been and are proving a financial success.

The Helen Mining Company owns a group of thirteen claims; the principal producers are the Confidence and Black Bird.   The greatest depth reached is 1,030 feet.   It is estimated that the production of this property has been approximately $1,000,000 up to January 1, 1904.

A 30-stamp mill, with 12 pans, 6 settlers and 9 concentrators has been erected at Graham on White-water creek about four miles from the mine where the ore is treated. A pipe-line

Fig. 23—CONCENTRATING MILL OF MOGOLLON GOLD AND COPPER COMPANY. Photographed by E. M. Chadbourne, 1904.

about three miles long runs up White water canyon and takes out the water from a large storage reservoir which is sufficient to run the mill machinery during the wet season. At such times when the water supply is inadequate steam is used. Electric drills are used at the mines. The property

has been idle since 1902, due to the death of the superintendent John T. Graham, who was one of the principal owners,
which occurred about that time

The Deep Down mine, was originally located by two prospectors, Bechtol and Foy. This property passed through
various hands and recently was organized by H. O. Bursum,
Governor M. A. Otero and A. R. Burkdoll, into the B. O. B.
Mining Company. This property has much to recommend
it, and has already produced $75,000, which came from near
the surface.

A 15-stamp mill with about 1,300 feet of work constitutes
the principal development. The ore is principally free milling gold and silver; however, baser ores have frequently
been encountered.

Lying just to the west of the Deep Down property is the
Maud S. group of three patented claims—the Maud S., the
Link and the Wilson. The Colonial Mining Company own
this property; the production has been $750,000. This prop-
erty has been idle for over two years.

The Last Chance is another noted property and its production is estimated at $250,000.

There is a 20-stamp cyanide mill on this property; owing
to some necessary improvements in the machinery the plant
was closed down during 1903.

Of the numerous producers of the Cooney district, the
Little Fannie group stands at the head of the list. The production is approximated at $1,250,000. Owing to litigation
the mine has been closed down for several years. A 15-
stamp pan-amalgamation mill is on the property.

Among other properties noted for their production and
development may be mentioned the Eberle group, Kat and
Kitten mine, Contention group, Grey Hawk, Copper Queen
group and Leap Year. Numerous other locations exist in
the district, some of which will no doubt prove valuable
when properly developed.

The total production of the Cooney district up to January
1, 1904, is approximately $4,650,000. The ores of the dis-
trict are gold, silver and copper; the principal production of
the camp, until recently, has been from the gold and silver

ores; pan amalgamation and cyaniding being the general methods of treatment.

The Mill of the Mogollon Gold and Copper Company is strictly one of concentration.

The Cooney mining district lies in the Mogollon mountains in the extreme western part of Socorro county about fifteen miles east of the Arizona line. Beginning near the western boundary of the territory, this range of mountains extends in nearly a north-south line for a distance of about ninety miles, diverging slightly from the Arizona line in passing to the south.

The western slope of the range is traversed by a number of deep canyons having their drainage in the Rio San Francisco.

On the eastern slopes the drainage is in the Gila river. The whole of this great water shed has been set aside by an Act of Congress as the Gila Forest Reserve.

It is observed that all the principal mines and mineralization is on the west flank of the range and not far from the foot of the main mountain slope.

There are three recognized mining districts in the Mogollon mountains, the Cooney, the Wilcox and the Tellurium.

The most important and northerly of these is the Cooney. Generally speaking, the formations are all of a volcanic type. The flows, in many instances, lie nearly horizontal, while others have been tilted until they approach perpendicularity. Andesites seem to be the predominating type of these volcanic rocks; while trachytes, rhyolites and basalts occur more sparingly. In many of the deeper canyons granite-porphyries are exposed, which would suggest that the basal granite is not far below.

There are two general systems of veins which cross each other nearly at right angles. The King and Queen lodes are the most prominent in the district; paralleling each other, they may be traced through a distance of ten miles. The Queen lode lies to the east of the King; the width of these two lodes varies from ten to thirty-five feet and each of them dips to the east.

The King lode, though a fissure, is properly a great faulted fractured zone, with numerous branches extending out from

it on either side.   These off-shoot or branch veins are the result due to the stress in shearing.

Much of the vein stuff is brecciated, altered country rock,
impregnated with the oxides of iron, carrying values in free
gold and cerargyrite.

The Queen lode is a true fissure as indicated by it cutting
through all formations and in its banded structure. The gangue matter is principally a combination of white and ribbon
quartz.   The dark grey bands of the ribbon quartz are rich
in argentite; the whiter quartz carry values in free gold.

The character of the veins lying between the parent lodes
vary through the forms of a contact shattered zone, contact
fissure, fault fissure and true fissure.

The Confidence and Last Chance vein cuts the King lode;
this with the Maud S., Little Fannie and Grand View lodes
are properly contact fissures. The principal production of
the camp is due to these contact leads.

A beautiful example of fault fissure occurrs in the Florida
and Golden Eagle lode; the displacement is, perhaps, 130 feet.
This fault line cuts the King lode at an angle of about 45ʾ;
its strike is approximately southeast and northwest.

Wilcox District.

This district lies to the southeast of Cooney about fifteen
miles, and was discovered in the fall of 1879 by William Wilcox, who was killed on April 29, 1880, the day before the
killing of J. C. Cooney.   There is no direct way of getting
across the mountains from Cooney to the Wilcox and Tel-
lurim districts, excepting by burro trail.   Generally the
trip is made from Cliff post-office; going about fifteen miles
to the north and bearing slightly to the west.

This district has been known a number of years, yet comparatively little development has been done.   The Zacaton
group is the most prominent of the locations.   Some ore was
shipped from this property about 1900, but the long haul to
Silver city consumed the profits.

The Silver Prize, I. S., Western Star, Uncle John and the
Butterfly group constitute the more important locations.

The character of the ore of the district is principally a si-
liceous sulphide carrying gold and silver.   An exception to

this is found in the Zacaton group which carries copper carbonate, in several of the locations. The Little Jessie has some very heavy iron oxide ores, through which chalcopyrite is disseminated.

In a general sense the formations are all eruptive in character and resemble those of the Cooney district.

Tellurium District.

Very little has ever been done to justify this selection as a distinct mining district.

It lies immediatly north of the Wilcox district about three miles, and has no distinctive features from that district.   It is said that some very rich pieces of Tellurium float were found there a number of years ago; hence, the name "Tellurium."

Tellurium No. 1, and No. 2 and the Pine hill claims owned by E. Peterson, of Mogollon, are the most promising prospects.

CHAPTER XIX.

MORENO AND WESTERN COLFAX COUNTY DISTRICTS.

The mining districts of Moreno, Willow Creek, Ute Creek, and Poñil can best be discussed in the aggregate, since they all lie on the immediate slopes of Elizabeth peak (Baldy) and their minerals and mineralization have their origin in that great isolated mountain. Elizabeth peak (Baldy), according to Lieutenant Wheeler in the United States Geographical Survey West of the 100th Meridian, reaches an altitude of 12,491 feet above sea level; this elevation is exceeded only in a few instances in the territory.

The writer has been very fortunate in having opportunities to familiarize himself, to a certain degree, with this section of the country. The beginning of this experience dates back to the fall of 1893, in making a railway survey, under the supervision of Ed. H. Smith, now of Las Vegas, from the Santa Fe tracks at Maxwell City, as far west as the valley of the Rio Grande. The line of this proposed railroad was through Cimarron, up the Cimarron canyon, through the gateway of the Moreno valley, where a branch line ran north from this point to Elizabethtown and into the Red River country. The main line, however, kept the Moreno valley south and crossed over the range at the Taos Pass, thence to Taos and from there down a boxed canyon to the Rio Grande.

Every facility was afforded to examine into the mineral resources during this survey. Topographical maps and geological sketches were made extending over a wide area of country, which embraced the mining districts flanking Elizabeth peak (Baldy mountain) and westward to Red river.

A geological section of the country, from a point east of Elizabeth peak to a point three miles west of Elizabethtown, was made at that time.

The geologic section best observed and studied is in the Cimarron canyon, that forms the narrow gorge and exposes the rocks to a thousand feet in depth on either side. The lower canyon, extending from near the town of Cimarron to Ute Creek park, is about twelve miles long and is entirely eroded in the thick bed of the Laramie coal measures; and which has, apparently, cut entirely through this latter series, since the Colorado shales of the Cretaceous are exposed at both ends of this part of the canyon. The upper canyon of the Cimarron, from Ute Creek park to the Moreno valley is approximately six miles in length, cutting through the Archean rocks and draining the great ancient lake of the Moreno valley. Recent development, it is understood, proves the existence of coal in the Moreno basin; its quality and extent is undetermined at the present writing.

In following up Ute Creek toward its source from its affluence with the Cimarron, it is observed that erosion of this stream is now in the Colorado shales.

At the west side of the Ute valley are two intrusive porphyritic members, which disappear toward the northwest as they pass into the eastern and steeper portion of Baldy mountain. Still further west is another porphyry member which has carried with it in its upward movement the Dakota sandstones.

On the east side of Ute valley is an escarpment of the light colored sandstones of the Laramie coal measures.

This sandstone may be traced on that side of Ute Creek until it reaches the saddle which divides the South Poñil from the latter stream. From this saddle the same series rises and covers the steep east slope of the mountain, even reaching and capping the very summit of the peak.

The sandstone, however, above the saddle to the top of the mountain is altered to massive quartzite.

The elevatory movement, which resulted in forming Elizabeth peak (Baldy), must have taken place at the beginning of the Tertiary. The origin of the mountain is due to the massive porphyritic sheets which have been forced upward between and through the thick beds of the Cretaceous. On the southwest side of Baldy in the vicinity of the Willow Creek placers, Cretaceous shales and limestones are superficially exposed. The cause of this exposure is due to the shattered

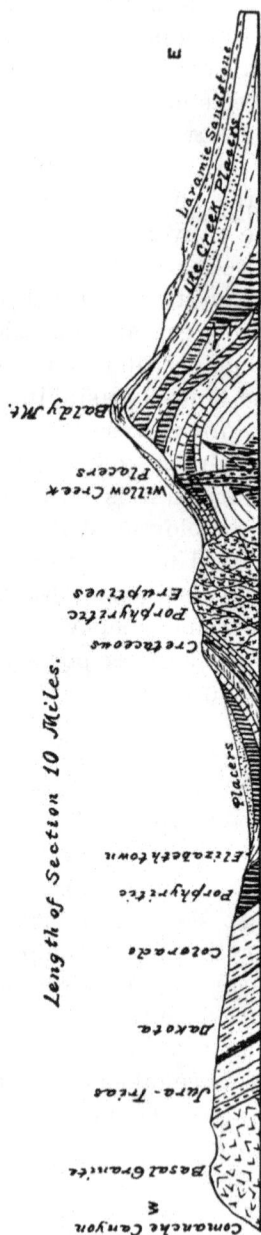

Fig. 24--GEOLOGICAL SECTION ACROSS ELIZABETH PEAK (BALDY) AND THE MORENO VALLEY TO A POINT THREE MILES WEST OF ELIZABETHTOWN.

condition of the strata that were elevated by the porphyritic intrusives and which favored more rapid erosion. In passing toward Elizabethtown from the Willow Creek placers, eruptives of a porphyritic nature occur immediately to the west of the gravel wash on Willow gulch.

Beyond and west of Elizabethtown are some exposures of the Permian and Jura-Trias. These latter series seem to rest immediately on the Archean core at Comanche canyon.

The whole of the area immediately flanking Elizabeth peak (Baldy) may be regarded as one vast placer field. The principal operations in placer mining have been conducted in the localities of Ute creek, lying on the southeast slope of Baldy; at Willow creek gulch, lying on the southwest slope; and the Moreno fields lying on the western slope in the vicinity of Elizabethtown.

These placer fields are the most extensive and have excelled in production those of any other placers in New Mexico. They are today producing over one quarter of the gold of the territory. To their success is due the fame of the districts.

Indirectly, the discovery of gold in this part of New Mexico is due to a Ute Indian who brought in a piece of rich copper float and exhibited it at Fort Union, during one of his trading expeditions, in the early part of the sixties of the past century.

The real discovery of gold in the Moreno valley, however, dates to October, 1866, when three men—Larry Bronson, ——— Kelly, and Peter Kinsinger—were sent by William Kroenig, W. H. Moore and others, from Fort Union to do assessment work on a copper property, near the top of Elizabeth peak (Baldy mountain), which had previously been located as a result of the find of the Ute Indian. This trio arrived on Willow creek late one afternoon and after arrangements for camping had been made, Kelly took a gold pan and commenced washing the gravel along the edge of the stream, while the other two prepared supper.

The first pan of dirt revealed several colors of gold. Calling to his companions and announcing his find, all three began to pan and dig, when to their astonishment gold was obtained from most every pan of dirt. Several days were spent in digging, running open cuts and panning with very satisfactory results. It being late in the season and the weather extremely cold, they gave up further prospecting until the following spring—forgetting, in the meantime, to do the work which they had set out to do, at the beginning of their mission. It was the intention of the discoverers to keep the find a secret after their return to Fort Union.

But secrecy could not be kept; the desire to exhibit the coarse gold saved by their pannings was irresistible. At the opening of spring in 1867, the excitement was so intense that the news of the find was spread over both New Mexico and Colorado. A great stampede was made to this Eldorado from all over the west. Bronson on returning to the diggings made several locations on the site of the original discovery, on Willow creek, for himself and partners. These locations were all measured from a big pine tree which was the initial monument, on which all courses and distances were based. This tree is known as the "discovery tree", and is now standing, April 1, 1904.

Numerous claims were staked during the first rush; among the new comers were Mathew Lynch and Tim Foley who made locations on the south or opposite side of the gulch. These latter gentlemen, especially Mr. Lynch, played an important part in the early history and development of the surrounding gold fields. To them is due the discovery of the famous Aztec mine, on the east side of the range. Mathew Lynch, with one or two minor exceptions, is to be regarded as the father of hydraulic mining in New Mexico. His efforts were very successful; running from three to four hydraulics each season, up to the time of his death, which occurred in 1880, by a tree falling on him.

About the time of the first locations on Willow creek, a second party from Fort Union arrived, and made the first discovery of gold at Elizabethtown. This party was composed of J. E. Codlin, Pat Lyons, Fred Phefer and Big Mich, who termed themselves the Michigan Company.

After the discovery at Elizabethtown, prospecting became general, and it was demonstrated that gold in paying quantities existed in most every gulch around Baldy mountain.

Tom Lothian, Dick Turpin and John G. Schumann were

the first to make locations in Grouse gulch. Across the Moreno river, just in front of Grouse gulch are the famous Spanish Bar diggings, which were located by Lowthian, Kinsinger and Bergmann.

Humbug gulch was located from the Moreno river, almost to its head. This latter gulch probably received its name from the supposition that it carried but little gold. Subsequent prospecting proved it to be the richest gulch in the district.

Fig. 25—MATHEW LYNCH, PIONEER HYDRAULIC MINER OF ELIZA-
BETHTOWN, AND THE ONLY SUCCESSFUL OPERATOR
OF THE "BIG DITCH."

Owing to the great influx of people into the new gold field, and in order to properly protect the rights of each individual, it became necessary to establish and organize a town. The prime movers in this direction were John Moore, George Buck, and a few others; the survey and plat were made by T. G. Rowe in 1867. Much discussion arose as to what the new town would be called; finally the name Elizabeth was decided on, in honor of Mr. Moore's eldest daughter, who is now

Mrs. Joseph Lowrey and is still a resident of the town which bears her name. As a matter of history, when Colfax county was first carved out of Mora county and organized, Elizabethtown became its first county seat; it also, bears the distinction of being the first incorporated town of New Mexico.

It was realized in the first years existence of the Moreno district that an ample water supply to work the new fields successfully was of prime importance.

Parties from Las Vegas and Fort Union took the water question in hand, and employed Capt. N. S. Davis, an Engineer of the U. S. Army, to make surveys and report on the feasibility of increasing the supply.

Surveys showed that a large supply of water could be brought around to the new fields by diverting the water from Red river, which was about eleven miles west of Elizabethtown.

Accordingly the Moreno Water and Mining Company was organized and L. B. Maxwell, William Kroenig, John Dold, W. H. Moore, V. S. Shelby, M. Bloomfield and Capt. N. S. Davis, composed the original members of the company.

The survey was finished and actual construction began on May 12, 1868. Work was pushed with all possible vigor up to November 13th, and the ditch was practically completed during that short period; as many as 420 men were on the work at one time. Considering the magnitude of the enterprise and the difficulties to overcome, this is regarded as the most remarkable engineering feat ever accomplished in the west. The flume in many places is suspended, for long distances, from perpendicular walls; deep arroyos were bridged and in one place the aqueduct crossed over a valley seventy-nine feet above the earth, requiring over 2,300 feet of trestle. The ditch from its head to Grouse gulch at Elizabethtown, is 41 miles and 660 feet long, with a carrying capacity of 600 miners-inches*, or 9,720,000 gallons every twenty-four hours.

---

*It should be noted that a miner's inch of water is not a definite or fixed quantity, but varies in different states and localities. Bowie's Treatise on Hydraulic Mining, page 268, states that in different counties in California, it ranges from 1.20 to 1.76 cubic feet per minute. In Montana it is 1.25 cubic feet. The writer has always taken 1.50 cubic feet, per minute, for a miners-inch in New Mexico.

A miners-inch in New Mexico may, therefore, be defined as that quantity of water which will flow through a vertical orifice one inch square, when the head on the center of the orifice is 6½ inches. By calculation it is found that water flowing under this condition, will discharge 1.53 cubic feet, during one minute of time.

One cubic foot of water equals 7.481 U. S. gallons and weighs 62.5 lbs. One U. S. gallon weighs 8.355 lbs. and contains 231 cubic inches.

The original cost was $210,000; an additional expenditure of $20,000 was made in constructing storage reservoirs at the headwaters of Red river.

Martin and Scott received the first water delivered by the ditch, on their property in Humbug gulch, July 9, 1869.

The Moreno Water and Mining Company constructed the ditch as a water speculation, and expected to sell water at such a figure as to pay a handsome dividend, since the company did not have any placers of its own to work.

The first water sold for 50 cents a miners-inch; owing to losses by leakage and evaporation and the great expense in maintenance, the company became financially embarrassed and the property passed into the hands of Col. V. S. Shelby of Santa Fe, who had loaned large sums of money used in the construction.

Shortly afterwards Shelby sold to L. B. Maxwell, the owner of the famous Maxwell land grant and one of the original members of the old water company. Mathew Lynch purchased the ditch from Maxwell in the summer of 1875.

Mr. Lynch operated the ditch successfully until his death in 1880; the property then passed into the hands of his two brothers, James and Patrick Lynch, who operated it for a number of years. This noted ditch is generally known as the "big ditch," though now it is practically inoperative from want of repairs; the probability is that it will never be used again as a water carrier for mining purposes.

In 1867, Joseph Lowrey came to the Moreno valley, where he works placers, by hydraulicing, just opposite Elizabethtown, at the present time; this ground is known as the Lowrey placers.

It is estimated that $2,250,000 of gold has come from the placer fields about Baldy mountain from the time of their discovery in 1866 to January 1, 1904; the average fineness of the gold is 885.

Owing to the angular form of most the gravels in the districts, adjacent to the foot of Elizabeth peak (Baldy), it is practically conclusive that the material has not been transported from any great distance. The theory has been advanced that no very rich lodes (with two or three exceptions) have ever been found in the mountain and that such enormous

Fig. 26.—HYDRAULIC MINING, LYNCH PLACERS.

quantities of gold as exist in the placer area must have come from some other source. The form of the gravels and the altitude at which gold is found on the mountain slopes, with the character of gold in the various districts, would preclude any theory, other than that the gold came out of Baldy mountain. The lodes about Baldy mountain were, perhaps, superficial in depth originally, as the Aztec and other rich veins indicate, and the enormous erosion has long since carried away the best mineralized part of the mountain and concentrated the value in the placers below.

Over on the east side of the range at the head waters of Ute creek is where the coarsest gold is found. This may easily be accounted for, since the Aztec lode is just above or at the point where Ute creek heads, and which lode is unquestionably the source of this coarse gold. One nugget* weighing a fraction less than twelve ounces was found a few years ago, just below the source of the creek. It should be remembered that the Aztec mine has produced many large nuggets of gold, found in the quartz, at the time the mine was worked. Over on the south Poñil from the Aztec are also found nuggets and coarse placer gold. The Aztec mine being on a ridge separating the two streams, would easily account for the gold found on both sides of the ridge.

Of the several producing districts and gulches of the Baldy mountain section, more gold is now being recovered from the Moreno river, just below Elizabethtown, than all of the other diggings combined. In fact, one-fourth of the gold of New Mexico is now being taken from this little stream by the Oro Dredging Company, under the successful management of R. J. Reiling.

This piece of machinery began operation on September 19, 1901, and is complete in every detail, being especially designed and constructed for working the gravels of this stream.

The material is elevated by a system of buckets and dumped into revolving screens where the larger stones are separated. As the coarse material passes through the screens it is washed thoroughly with jets of water, and the mass of

---

*This nugget forms a part of the collection of gold nuggets of the Maxwell Land Grant Company.

finer gravels sand and mud is forced by a heavy torrent of
water into a long iron flume or sluice box.   In this sluice box
is placed a suitable false bottom and several pounds of
quicksilver are distributed in the riffles, in a manner similar
to that of the ordinary wooden sluice box.

Fig. 27—DREDGE OF THE ORO DREDGING COMPANY, OPERATING
IN THE MORENO RIVER AT ELIZABETHTOWN.

The process after the material is elevated, is identically
the same as is practiced in ordinary hydraulic sluicing; that
is, forming riffles in the sluice and amalgamating by the use
of liquid mercury.   Clean-ups are made about once every
week and the product is retorted in the usual manner.

The capacity of this dredge is 4,500 cubic yards per 24 hours; although the average daily amount handled is about 3,000 cubic yards, This machine can successfully work a gravel bank 25 feet in thickness and can excavate with ease 15 feet below the surface of the water.

Next to this dredge in production is the hydraulic plant of Joseph Lowrey, who has been operating every season for over thirty years.

Other placers in the Monero district, beside those of the Oro Dredging Company and Joseph Lowrey, are worked in a small way. The Lynch Brothers have not operated placers for several years.

In Willow creek district, the more prominent grounds are the Kaiser diggings, Last Chance and Grub Flat placers, and the Brown diggings.

On the Ute creek side are the Pierson and Mead diggings and the Major Dennison placers.

Across the south Poñil are the Wallace placers. Among the phenomenal gold strikes in New Mexico, the Aztec lode deserves special mention.

Mathew Lynch and Tim Foley, after making some locations in Willow gulch, crossed over the range to the east side, where they found rich float and coarse gold at or near the head waters of Ute creek; this was in the summer of 1867. These two gentlemen worked assiduously for about one year in tracing the rich float to its source. Finally in June, 1868, they uncovered the Aztec lode. This was said to have been the richest discovery made in the west, of gold from a vein or lode. As previously mentioned, this mine is at the foot of Baldy mountain on a little ridge which separates Ute creek from south Poñil. The Aztec lode is a contact between shale-slate and quartzite. In other words its geological horizon is in the cretaceous shales which immediately underlie the Laramie coal measures. This is, perhaps, the most singular occurrence of gold, when considered from a geological stand-point, of anything in the west.

The strike of this vein at the surface is about N. 40° W. and dips to the northeast about 75°, to a depth of 90 feet.

Below the 90-foot level the dip almost attains horizontality and conforms closely to the contour of the hill. The strike also changes to about N. 70° W.

The original ore was of a brown color, due to the oxidation of pyrites; the gold being remarkably free milling.

A 15-stamp mill was immediately erected and operations began on the 29th day of October, 1868.

The mine was successfully operated between three and four years time, and was closed down in 1872 on account of litigation. The principal rich ore bodies were, however, worked out during that period. The mine since that time has been spasmodically worked, though never successfully in a general sense. The production to date is estimated at about $1,250,000; about $1,000,000 of this was mined prior to 1872.

Many promising lode claims exist in the different districts which surround Elizabeth peak (Baldy). In the Poñil district is the noted French Henry mine; it being the first claim located in this district. Among others are the Smuggler, Guerilla, Mountain Witch, Paymaster group, Black Joe, Harry Bluff, Harry Lyons group and the Mount Vernon.

In the Ute creek district may be mentioned some of the more prominent lodes, as the Thelma, Montezuma, Black Horse group, Rebel Chief group, Maid of Erin, Rosita, Puzzler, Monarch, Homestake, Bull-of-the-Woods, Paragon, Little Jessie, Sweepstakes and Real.

The Willow creek district has several producing and promising lode properties such as: The Golden Ajax, Legal Tender, Hidden Treasure, Golden Dollar, Ophir, Only Chance, North and East Pacific, Grand View, Mystic, Victor, Indiana, Alababa, Little Wonder, Grand Duchess, Mark Twain group and Aristocrat.

The most important group of lode claims in the Moreno district is the Red Bandana. The names of the leads composing the group are the Red Bandana, Empire, Moreno, Centennial, Galena, and American Flag.

This property is extensively developed with over 2,000 feet of work, with 5-foot Huntington mill, engines, boilers and shaft-house. Since the property lies on the ridge which separates Grouse gulch from Humbug gulch, it is quite evident that the Red Bandana lodes contributed a very great part of the gold found in those gulches. The formation consists of slate, quartzite, porphyry, granite, serpentine and limestone; the strike of this series being east and west. The veins of the group run approximately northeast and southwest, cutting all formations; hence, they would be of the true fissure type. Near the surface the ore is free milling; but changes to a sulphide at depth. Other prominent properties are the Abraham Lincoln, Heart-of-the-World, Iron Mask, North Star, The Baldy Mountain Tunnel group, Bob-ta l Senate, Penuckle group, Imperial No. 2, Sheridan, Golden Era, Gold Leaf, and Admiral Dewey in Big Nigger gulch.

West Moreno (Hematite) District.

Lying northwest of Elizabethtown about five miles and to the left of the road in going toward Red river, is the west Moreno or Hematite district. This district lies in the extreme western part of Colfax county and took its name from the highly colored formations and veins, due to the red oxide of iron. The Archean nucleus is well represented along the lower part of Hematite creek, having an exposure of more than a mile in width. Many mining claims have been staked out along this gulch and especially further to the west, where the country is crossed by a parallel belt of porphyritic eruptives, and highly altered slates. Some placer gold is found along the gulches, though not enough to justify profitable working.

Development has not been very extensive; the ores being refractory and usually low grade, the excitement which attended the first discoveries has now died out completely. The camp was practically deserted during the writer's visit in June, 1903; notwithstandng, a number of very favorable prospects are kept alive by outsiders who do their annual assessments.

The most favorable properties are the Black Wizard group, Iron Bird, Challenge, Kentucky, Last Chance and Gold Belle.

Hematite creek flows east into West fork and thus aids in forming the west prong of Moreno river.

Urraca and Bonito District.

This district is in Colfax county and is situated about ten miles southwest of Cimarron. During the summer of 1896, a man by the name of Craig found some ore that gave a very high assay value, but nothing further has been heard. Some placer gold is found in the gulches in the vicinity of Urraca creek, which may eventually be worked.

Cimarroncito District.

This is substantially the same as that of Urraca and Bonito, as above mentioned. Very little attention has been given this section, except in a very superficial way.

The country is considerably broken by eruptive dikes and porphyritic intrusives, which have induced mineralization in a more or less degree in certain localities. Some veins, apparently of a fissure type, seem to have the most favorable showing.

Among the latter are the Mocking Bird group and the Big Missouri No. 2.

It would be well to note that all the mining districts of this chapter are embraced within the limits of the Beaubien and Miranda (Maxwell) grant.

CHAPTER XX.

RED RIVER DISTRICT.

This district lies in the eastern part of Taos county, near the line of Colfax county, about twelve miles northwest of Elizabethtown. The country up and down Red river with many of the tributary gulches was prospected in 1869, about three years after the discovery of gold at Elizabethtown. Some placer gold was taken from the stream and gulches; though not in the quantities found about Baldy mountain at Elizabethtown.

In 1879 the Waterberry Company built a smelter on the property now known as the Copper King.

On account of a deficiency in fluxing materials, and the great distance from a railroad, La Veta, Colorado, being the nearest railway point, the enterprise proved a failure. This smelter was never operated after a trial run or two, and burned down in 1889.

Beginning with the years of 1893-4, the first systematic prospecting was inaugurated. It was during the year 1894 when the present townsite of Red River City was located by the Mallette brothers.

During the fall of 1895 the Golden Treasure and Jay-hawk were located.

About the time of the permanent settlement of the town, Anchor or Midnight was settled, and two mills at that place were operated for a while; litigation closed the mills and they are at present idle.

Red River is a beautiful mountain town and is fast gaining reputation as a summer resort.

No mining camp was ever favored with better facilities for wood and water.

The stream of Red river (Rio Colorado) is an ideal one for the installation of electrical power for the purpose of operating all kinds of mills and mining machinery. The time in the future is not far distant when the "eternal hills" will open up their treasures by responding to the magic touch of a button. The millions of horse-power now going to waste here annually, is ample to reduce the ores of the district and convert the rusty metallic sulphides into shining gold.

This district is not lacking in the series of rocks, such as granites, gneisses, syenites, schists and quartzites. Some shales, slates and limestone are found flanking the older formations; the latter, especially, are not very conspicuous. The whole district is checkered with intrusives of diorite and other eruptive types of rocks; the surface is frequently capped with andesite.

Porphyry dikes carrying iron and copper are numerous; the influence of such dikes in the segregation of ores is quite marked in many instances; the Jay-hawk and Copper Dome furnish good examples in this respect.

Some of the more prominent claims and prospects in the Red River district are: Black Diamond group, Peerless group, Homestake group, Copper Dome, Anaconda group, Standard group, Laura B. and Minnie L., Last Chance, Fort Reno and Deadhead group, Yankee Maid, J. O. G., Deadwood group, Ragged Pants Dick, Golden Treasure, Paxton group, Copper Hill group, Hornet, Rock of Ages, Sure-thing group, Exile group, Bueno, Commodore group and Wild Rose group.

The Jay-hawk mine is the most noted property in the district; H. J. Luce is manager. This mine is situated about three miles above Red River City on what is known as Black mountain. The property consists of five claims; three of which are patented. Present development (January, 1904) consists of over 300 feet of tunnel and 50 feet of shaft. Five distinct leads have been pierced by the tunnel; one of which is nearly 22 feet wide, another 10, and the last one is small, but carries the best values. At present this ore is hauled down the mountain, about five miles, where it is concentrated. The capacity of the concentrator is about 30 tons daily. Two Wilfley tables, a sixty horse-power boiler, a fifty horse-power Corliss engine, and a crusher form the principal equipment. In order to avoid the long haul of ore to the mill, the company will soon have a new mill completed which is being erected at the mine.

Beside the concentrating plant of the Jay-hawk company, there are three arrastras being operated in the district with fairly good success; others to be installed soon.

This district has many good features; the only thing disparaging, is the long distance from transportation.

On June 18, 1904, John Lacaniche and Louie Marchino made an exceedingly rick strike on their Independence mine which suddenly revived the camp. This particular property is on Bitter creek, about six miles from Red river.

Considerable excitement prevails at this writing and the country is being covered with locations.

The classification of the ore has not been fully determined; it is thought to be petzite. The Angola Boy was the next and adjoining claim to the Independence that was staked; it belongs to F. C. Stevens, L. Rogers and F. L. Stevens.

Black Copper District.

This district, properly speaking is not a distinct district, but belongs to the Red River district. It is sometimes spoken of as the "head of the river," having reference to the head or source of the Red River.

Some very rich ore was taken from the original Black Copper mines a few years ago; extensive development was done then and a ten stamp mill with concentrating machinery was installed. Gold was the principal metal of the ore. Litigation and other matters caused the property to close down; it has now been idle for several years.

Theodore Cannard owns several promising properties here of which mention was made under the Peerless group of Red River.

This camp, unfortunately, is very inaccessible.

Keystone and Midnight Districts.

This region lies immediately to the north of Red River district and is in Taos county. The general character of the ore and formations are similar to those at Red River. Development has not been very extensive; although at one time considerable activity was manifested in the camp.

The only work of importance carried on at this writing is by the Cashier Gold Mining and Milling Company, in the development of the Cashier group of claims.

It is probable this property will be a producer during 1904, since present developments are very encouraging.

La Belle District.

Considerable excitement existed in this district about the year 1895. The region is very similar in most respects to the general country in the Sangre de Cristo range of mountains.

Porphyritic eruptives and intrusives form the principal features to be seen on the surface; the veins are principally fissures.

During the time of the first excitement a large number of claims were staked out; most of which, however, fell back to the government a few years later.

Of the old locations, the Aztec is favorably known, which belongs to the Aztec Mining and Milling Company. It has a shaft 100 feet deep and a tunnel 176 feet deep.

Generally speaking, the ores of the La Belle district are low grade and very refractory. Transportation is badly needed.

Rio Hondo (Twining) District.

The Taos range of mountains in New Mexico is to be regarded as only a continuation of the Sangre de Cristo range which lies farther to the north in Colorado.

Colfax and Taos counties mutually share the backbone of this elevated ridge as their common boundary line.

The rugged formations of the region are primarily due to volcanic forces, aided by finishing touches of erosion, effectually applied in later geological epochs. Lying high on the western slope of the range whose drainage waters at this point reach the Rio Grande, by way of the Rio Hondo, is the Rio Hondo mining district. Mining operations have been carried on in this district, in a small way, for a great many years.

Much work in placer mining was formerly conducted by the Mexicans near or at the foot of the mountains. At Amizette many years ago, a small copper smelter was erected which proved a failure.

It was not however, until early in the spring of 1893, that the district received something of a boom by the discovery of a number of promising mining claims in the vicinity of the old camp at Amizette.

The writer was among the earliest visitors to the new fields, and had an opportunity to familiarize himself with the district to a considerable degree; having remained there during the months of May, June and July.

The district is very rugged; the main axis of the range extends north and south and has a gneissoid and schistose structure. Intrusive dikes of diorite and porphyrite cut the schist formation at intervals; some of which are iron bearing and frequently carry a good grade of hematite.

The ore lies at the porphyry contact and also in and through the schist; the latter formation being interspersed with seams and veins of quartz. The field term for these schists may be termed hydro-mica and are perhaps Algonkian in age. The igneous action of these intrusives on the schists and shales accompanied by hot waters, is the primary cause of inducing mineralization.

The Fraser mine is by far the most important in the district. There are several claims in this property, some of which have been extensively developed. The property lies well up toward the top of Fraser mountain; a tram of about half a mile in length conveys the ore to the mill.

A one hundred ton concentrating plant and smelter has been erected by the Fraser Mountain Copper Company and some trial runs made. Since the ore is of a very low grade, averaging less than $5.00 per ton in copper, gold and silver, the enterprise was not a success. The writer was on the ground about the time the first run was made; from observations made during that visit, it was not surprising to learn, shortly afterward, that the concern had gone into the hands of a receiver.

The mines should not be held responsible for this misfortune; the blame should be attributed solely to incompetent management.

Bull-of-the-Woods group lying partly in the Red River district, is rather a favorable looking property; it lies somewhat across the divide east of the Fraser mountain. Development to the extent of a 500-foot tunnel and a 65-foot shaft has been done by the owners, James Lynch and associates of Elizabethtown. This property is a copper proposition; the ore is found in a schistose rock carrying some quartz. On account of its inaccessibility it has been lying idle. A very extensive low grade cyaniding proposition is found in South Fork canyon and known as the South Fork group; it is being developed by the San Cristobal Copper Co., of New York.

The King Solomon group, Berry Extension and Copper King group, including some free gold claims lying in Long's canyon, embrace the remainder of the principal lodes; very little development work has been done on these properties.

Near the top of the divide in going over the trail from Twining to Red River, above the Fraser mill, is found some placer ground which was worked in the fall of 1893 by hydraulicing, but the enterprise was only partially successful; the water supply being inadequate on account of the position at the top of the range. Nothing since has been done at these diggings, only in a small way by panning.

Cieneguilla (Glen-Woody) District.

Nowhere in the southwest is such an enormous body of quartz and quartzite to be seen as at the Glen-Woody camp in the Cieneguilla mining district.

This camp was established in the summer of 1902, by W. M. Woody, who represents the Glen-Woody Mining and Milling Company.

Mr. Woody, before going to Alaska at the time of the Klondike gold discovery, operated placer mines on the Rio Grande, one half mile below the present camp; this was the time when his attention was first drawn to the immense body of quartzite.

These placers were in the gravel bed of the river; but were not successful financially, owing to interference of *mal pais* boulders lying in the bed of the stream. It was at this time that attention was attracted by the enormous quartz deposits which were exposed along the stream.

Glen-Woody camp is where latitude 36° 20′ North, crosses the Rio Grande, in Taos county. The townsite is on the west side of the river, while the mill and ore body is on the opposite or east side.

The bridge at this point was built on some of the old piers

of a government bridge which was burned by the Apache Indians during their hostilities in the latter seventies; the bridge was never rebuilt by the government.

An experimental plant of 50 tons capacity, consisting of a a 5-foot Huntington mill, with cyanide tanks has recently been completed.

The machinery is run by water power; the water being conducted by a flume out of the river about one mile above the mill. The power is developed by a 160 H. P. turbine wheel. This power is ample to run several more Huntington's should the experimental runs be a success.

The Glen-Woody ore may, generally speaking, be classed as quartzite; although stringers and veins of quartz run through the deposit at numerous places. These quartz veins and stringers sometimes attain a width of several feet. The width of the main lode itself is approximately 600 feet; and rises to an average height above the river of 300 feet. The Company has three full claims, making 4,500 feet along the lode. Approximately, there are 50,000,000 tons of ore above the river which may be considered in sight. The length of this enormous lode cannot well be ascertained, since it crosses the river about half a mile below the bridge where it is covered by the mesa lava, and in following up stream the river deflects to the left gradually leaving the lode more than a thousand feet at the easterly end of the Glen-Woody property.

It is quite probable that the lode is four miles long; the northeasterly end disappearing under detritus at the top of the hill.

From a survey made by the writer of the property in November, 1903, the strike of the lode was found to be N. 60° 22′ E. and the dip about 80° to the southeast.

A much sheared quartzite lies on the foot wall next to the river, and a similar quartzite is on the hanging wall though of a lighter color. The exact age of this formation has not been determined; though it probably is Cambrian.

The general appearance of the main lode is of a reddish tint; although some of the best grade of ore is a granular white quartz. The only question concerning this property as a merited mining proposition, so far as the writer is able

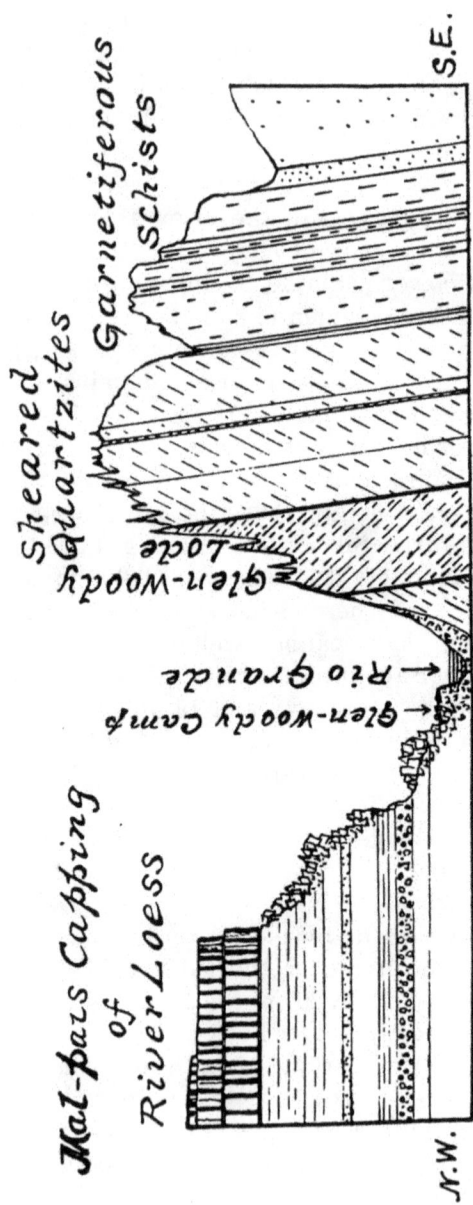

Mal-pais Capping
of
River Loess

Sheared
Quartzites

Garnetiferous
Schists

Glen-woody
Lode

←Rio Grande

←Glen-woody Camp

N.W.

S.E.

Fig. 28—GEOLOGICAL CROSS-SECTION OF THE GLEN-WOODY LODE.

to judge, lies in the value of the ore, which is claimed by the proprietors to average $2.50 per ton in gold. Should this ore only average $1.50 per ton, there can be no reason why the enterprise should not pay a handsome profit; since it is an ideal cyaniding proposition and every condition is favorable, such as cheap labor and power, water, climate and transportation.

Copper Mountain District.

Lying a few miles to the east from Glen-Woody camp is the Copper Hill district, which is practically the Picuris region. This district has been recognized for a number of years by its characteristic outcroppings of strong leads, though little development has ever been done; the principal exception being the work accomplished by the Copper Hill Mining Company. This company did its development in 1900; having run a 600-foot tunnel and sunk a main shaft to a depth of 180 feet.

A hundred-ton concentrating plant was erected at this time, but unfortunately burned after it first started. This circumstance forced the company into the hands of a receiver and the property was recently sold to satisfy certain judgments.

Within the past two years the Green Mountain Copper Company, operating at Green mountain, has done some development. A tunnel of 310 feet has been driven into the mountain to cut the main vein at a considerable depth below the surface. The ore is of silicious character carrying copper, gold and silver in sufficient quantities to pay if the property is intelligently handled. The vein is a contact between quartzite and mica-schist. The general make-up of the district is similar in many respects to the Glen-Woody camp— quartzites and mica-schists being the parallel features.

The Ural group is fairly well developed, as also, the Wilson group; the chance of each of these properties becoming producers is good.

CHAPTER XXI.

BROMIDE DISTRICT, No. 2.

This mining district is situated in Rio Arriba county, about fourteen miles west from Tres Piedras, a small village on the Denver & Rio Grande Railroad.

The basal formation is granite; lying next to this core are the gneisses and heavy Algonkian schists. Flanking the older formations are the sedimentary series all of which are more or less shattered and covered by eruptives of trachoryte (Hayden).

The whole country about this point, from the northeast around to the southeast is covered with basalt *(mal pais,)* to the Rio Grande; and even beyond that stream the flow extends for several miles. This great lava flow dammed the river for a distance of sixty miles; the evidence of this statement is verified from the nature of the great gorge which is over 1,000 feet deep and less than half a mile wide.

Tusas Peak is about 9,500 feet in elevation and is the highest point in the camp; the summit of which is composed of the older indurated rocks which have been exposed to erosive agencies for countless centuries.

An axial ridge from this peak runs in a flat V-shape; the vertex of the letter very nearly corresponds with Tusas Peak: and the sides of the V take the directions of the axial ridge, running northwest and northeast, respectively.

The principal mines of the district lie on either side and near to the left branch of this V-shaped ridge.

The whole country is considerably faulted and the district is noted for its vast bodies of schist. Occasionally, lenticular bodies of quartz are also encountered. Sheared quartzite in juxtaposition with the schistose formations and gneissoid granites would indicate tremendous slipping pressure, which was certainly accompanied by great "heat, and in the presence of hot waters, the metallic sulphide ores were segregated along the most favored zones of fracture and porosity. All of the ores, generally speaking, are sulphides; copper, silver, lead and pyrites of iron carrying gold. Copper, associated with gold, will no doubt be the principal metal found in the deeper levels.

The first lode discovered and located in this district was the "Bromide," from which the district took its name.

The property was located in July, 1881, by D. M. Field and J. M. Bonnett. The ore is more of a brittle silver (stephanite) rather than the bromide (bromyrite).

The property has been little more than merely prospected; about $18,000 has been taken from the mine, some of the ore being exceedingly rich. The ore is purely silver, no gold nor copper being present.

Work on the Dillon Developing Tunnel has been recently started at the southeast corner of the Bromide claim, which will pass under the old workings of this mine at increased depth. This tunnel will then be slightly deflected and driven under the Pay Roll mine, giving 760 feet in depth on that property, and draining a number of other properties beside the Bromide and Pay Roll. The length of this tunnel will be 6,300 feet, and will cut at right angles all the heavy schist formations. The completion of this tunnel would fairly demonstrate the value of the schist belt in that section of the camp.

Among the more prominent properties of the camp is the Admiral group, discovered by Thomas Smith, a mining engineer, about 1902, which consists of five claims; the three claims—Dewey, Sampson and Schley—lie on the north side of the axial ridge, while the Wedge and Blue Bell lie on the south of the ridge. The top of the axial ridge virtually forms the boundary line between the two latter claims and the three former ones.

An immense dike of trachoryte (Hayden) traverses these claims and extends six or seven miles; the walls of the dike are gneissoid in character, and are from 800 to 1,500 feet apart.

Contact veins exist along either side of this eruptive having the gneiss for a wall of each vein, with a schist capping.

The general character of the formation of most of the claims in the district is similar to that of the Admiral group.

Fig. 29.—A VEIN OF THE DANBERRY MINE.

Some splendid ore, copper carbonates and sulphides carrying gold and silver, has been taken out of the Tampa quite recently; the indications are that the ore body is permanent. This property belongs to the Tusas Peak Gold and Copper Mining Company. Other properties, such as the Whale, Mayflower group, 16 to 1, Sardine, War Eagle, Mexican King Co., Merrimac group, Midnight, The Wayne-Arriba Mining Company, Last Dollar, Keystone Co., Pontiac Mining Co., Walker, Iron Clad, Gold Pan Co., Butterfly, Red Fissure, Farragut, Agnes, Royal Purple, Independence, Strawberry, Joe D., Big Sandy, etc., deserve mention. It is learned that several good prospects have been opened up on Cow creek which lie nearer the railroad by three or four miles. The Bromide district is well watered and heavily timbered. Whilst the district has never been a heavy producer, this should not militate against the camp; since the title had been under a land grant cloud until 1900, when the United States Supreme Court declared the district public domain. Since that time much progress has been made; development is partaking of a substantial nature.

Headstone (Hopewell) District.

Originally the Headstone district embraced that of the Bromide. The Bromide district was organized and established by cutting off and comprising a certain portion of the original Headstone district.

During the summer of 1881, after the rich strike made in the Bromide, a wagon road was built into Hopewell (then Good Hope), on account of rich placer discoveries at Eureka gulch about a mile out from Hopewell.

It is said that a merchant by the name of Clarence C. Hood took in about $175,000 of placer gold which came from that gulch during the first three years he was there. The largest nugget found at that time was valued at $34.00. Mr. J. P. Gill has just completed (December, 1903) a very fine hydraulic plant at the Lower Flat placers, west of Hopewell, when he expects to turn on the water, early in the spring, and begin active work at that time. These grounds are gulch gravels and are said to be very rich.

The formations here are substantially the same as those at

Bromide. Granite, gneiss, schists, and porphyry prevail; the ore being found on the contact between the granites or schists, or wholly in the schists by penetrating quartz veins and stringers. The ores in this camp are principally sulphides and carry gold.

The largest nugget of gold ever taken from any of the Hopewell lode properties came from the Croesus, at a depth of forty feet and was worth about $15.00. This would seem that the origin of the Eureka and Lower Flat placers is from the adjacent lodes. The extent of this placer area is not considered great.

The lode claim known as Mineral Point, and worked under the management of E. C. Sterling is rather extensively developed; having something like 1,500 feet of work.

The ore is apparently low grade and refractory and would average perhaps, $9.00 per ton. A little copper and silver is also found in the ore.

Entensive work is now going on in the "Jaw Bone:" diamond drills have been used in the prospecting. Lately the drills have been laid aside and work in extending a tunnel and sinking a shaft is being prosecuted; this would indicate that the diamond drill had pierced some good ore.

Some of the claims of this district are the Good Hope group, Crescent, Duck group, Golden Age, Atlantic Mining Co., Silent Friend group, Hornet group, Iron Mountain, Columbia group, 10 Better group, Buckborn, Hidden Treasure group, Emerald group, and others.

The same conditions prevail in this camp as those found in the Bromide; active work on a business mining basis is now being inaugurated, and good results are expected in the near future.

Copper Canyon District.

This cannot with propriety be called a mining district, since practically no prospecting or development has been done. It lies near Abiquiu, in Rio Arriba county.

The principal location is the "Lily Belle" owned by J. E. Irvine. The work consists of a 60-foot tunnel. Copper glance is the character of the ore and it occurs in a white sandstone.

Ojo Caliente District.

This district received its name from the famous hot springs in that locality. But little mining has been carried on in this region. The most important property in the district is the mica mine, known as the Mica Age, belonging to Hon. Antonio Joseph and mentioned under the chapter on mica. The Antonio Joseph group of gold and silver mines lying a short distance west of the hot springs is the best developed in the district. It is thought by some that this property was worked by the Spaniards, as there seems to be some evidence in one place of ancient working. The Chicago and Big Missouri are the only others deserving mention at this writing.

The region has been developed in the throes of volcanic action and is much torn and shattered, and the fires of its more youthful days have not yet cooled, as evidenced by the existing hot springs.

CHAPTER XXII.

WHITE MOUNTAIN DISTRICT.

In the early days, prior to 1870, the whole of the country stretching from the White mountains to the Jicarilla was known as the Sierra Blanca region.

Many old ruins were noticed at the time of the first prospecting, of which but little or nothing is known of the former inhabitants. Traditions among the Indians and Mexicans have it that those ancient people worked rich mines of gold and silver somewhere in the Sierra Blanca, the localities of which are at the present time unknown.

Prospecting for placer gold was carried on in a desultory manner by Mexicans in this region, as early as 1860.

Among the oldest claims in the White mountain region is the Sierra Blanca lode; this was located about the year 1868. The vein is encased in what appears to be a kaolinzed augite-andesite; and seems to have a width of nearly thirty feet. Gold is the principal metallic value of the district; no active mining is being conducted at present.

Nogal District.

Nogal district lies in Lincoln county and the country is rendered conspicuous by the celebrated Nogal peak which rises to the magnificent height of 9,983 feet above the sea.

The first authentic account of the discovery of gold was in Dry gulch which heads above the American and Helen Rae mines; this was about the year 1865. Billy Gill is credited with being the discoverer of the first quartz lode, called the American mine; this was in 1868. Operations on this lode were conducted from Ft. Stanton, presumably by soldiers.

A number of the early prospectors, at the hands of the Apache Indians, laid down their lives in this region in pursuit of the seductive yellow metal; two of the unfortunates were buried where they fell, near their prospect pits.

No systematic prospecting was done in the Nogal district,

until the year 1880; the district at that time formed a portion of the Mescalero Indian reservation, which was, in part, thrown open in May, 1882. The following October the Rockford mine was opened and some very rich ore, running $80 per ton, was encountered.

M. M. Gaylord and Company in 1883 erected the Rockford mill; cleaning up $8,000 in a sixty-day run. Litigation followed and the property has remained practically inoperative ever since. The Rockford and two other claims, known as the Clipper and Cashier, constitute the Rockford group.

Milling of ore from the American mine started in October, 1884, by Gill, in connection with Joseph Fletcher who owned a mill.

The mill was equipped with Cornish rolls, and in eleven months time $23,000 was produced by only two miners working in the mine.

In 1885 Godfrey Hunter purchased the property for $11,-000; a little later disposing of two-thirds his interest for $25,-000. New York parties afterward acquired the property, but never were able to make the proposition pay. The production of the mine, all told, is estimated at $85,000, to January 1904.

No claims in the Nogal district are better known than the Helen Rae and Cross-cut; located by three "tenderfeet," R. D. and Harry George and Chas. Epps, in 1880. Much of the ore from these two claims was phenomenally rich which was taken out by surface trenching along the lode and by shallow shafts. After taking out considerable gold in 1880, the three partners, the following year, from some cause failed to do their assessment, and two prospectors by the name of Storms and Murphy in 1882 jumped the property. A little later these two men sold out to John Rae, from which name the property is now known. Rae pounded out with mortar and pestle from $100 to $300 per day from the Cross-cut; and in one instance one single pan of the ore yielded $1,100 in gold. A shaft sunk on the Helen Rae twenty-five feet deep gave $3,-300; $10,000 was mortared and panned out, and 125 tons milled $4,804, making a total of $14,804 taken out by Rae in less than one year. The ores turned base at no great depth; this consequence induced Rae to sell the property, which he did to Rolla Wells for $15,000. Wells in turn passed the mines over to M. D. Gaylord in 1901, who is connected with the American Gold Mining Company, and who now in 1904, operates the property. The Ibex is also a property with much to commend it.

The strike of the veins in this part of the district is nearly north and south and the dip is about 80° toward the west. Formation is andesite, syenite and birds-eye porphyry, with veins cutting through; kaolinization has advanced to a considerable degree.

Nogal village, which owes its birth to the mining in the early 80's in that region, is located at the junction of Dry gulch and Nogal canyon, north of the Helen Rae and American mines.

Bonito (Parsons) District.

Over the divide from Nogal toward the south, a few miles, the Bonito mining district is situated, and lies in Lincoln county.

Near the divide which separates the two districts, old workings, presumably of Spanish origin, are to be seen; turquoise appears to have been the object of their efforts. Kaolinization has taken place and the formation is almost a counterpart of those at the Cerrillos and Burro mountain turquoise mines; some fairly good turquoise fragments were picked up.

The prevailing types of rocks in this district are much the same as on the Nogal side of the mountains; birds-eye porphyry and syenite are the chief classes.

The most noted mine in the Bonito district is the old Parsons' (now Hopeful) mine. This property at one time had a considerable production. The postoffice of Parsons is about one mile below the mine.

The main lodes have a general north and south strike; while the subsidiary and more recent lodes cut through the older system nearly at right angles. Gold predominates in the primary lodes, while lead-silver is the distinguishing characteristic of the later or cross lodes. On the whole, the various classes of ore are very low grade.

A few of the locations are the Etta, Bismarck, No. 1 and

No 2, Lady Francis, Jennie Lind, Buckshot, George Washington, Martha Washington and the Crow and Raven gruops.

Among the early prospectors and pioneers of Bonito district are C. C. Boürne, R. C. Parsons, Col. G. W. Prichard, Ed. Beard and others, who came into the region in 1881.

Fig. 30 -SCENE ON RIO BONITO.   Photographed by F. A. Jones, 1904.

The old Bonito camp or postoffice is now practically abandoned, only two persons linger in that once lively little mountain village.

Eagle Creek and Rio Ruidoso Districts.

Southeast of Bonito a few miles, and a little west of south from Fort Stanton are these isolated districts; little or no

production has ever been credited them. This region lies in
Lincoln county, and was first prospected in 1885 by John A.
Holder, John C. Hightower, Capt. Brazell, Charles Berg and
Sam Dougherty. Porphyry and granite are the prevailing
types of rocks at the center of the districts, flanked by limestone toward the outer edges. Copper and silver ores appear to predominate; some gold and lead ores are encountered occasionally.

The veins are principally contact fissures and have a general north and south trend.

Principal among the prospects are the Comanche, Chance,
Return, Virginia and Modoc.

White Oaks District.

Of New Mexican mining camps none, perhaps, is better
known than White Oaks.

It seems that the Mexicans had washed gold in a small way
in the lower part of Baxter gulch, as at Jicarilla, for a number of years prior to the rich discoveries which made the
camp famous.

Not until the fall of 1879 was the news heralded abroad
that the discovery of fabulously rich gold ore had been made
on Baxter mountain.

A number of prospectors had been prospecting the immediate vicinity for placer gold, among them were George Wilson and his partner Old Jack Winters, and George Baxter.
While the party were eating dinner, Wilson took his lunch in
his hand and strolled up the side of Baxter mountain, where
he climbed on top of a large "blow-out" and with his pick
chipped off a piece of the rock, and on examination was much
surprised to find that it contained gold. He immediately reported his find to those below and staked out the North Homestake, which was the first lode location made in the camp.
On this same afternoon Wilson relinquished his rights in the
property to his partner, Jack Winters, for $40, a pony and a
bottle of whisky. Not a great while after this deal, the discoverer of the lode mines of White Oaks disappeared and
was never heard of again.

Winters who became the sole owner of the Homestake
shortly afterward died, leaving the property to his heirs,

who sold their interests to James M. Seigfurst for $50,000. This gentleman, during the first year took out enough gold to pay for the mine, the building of a mill and had a surplus of $10,000. It is said that $35,000 was taken out in two days; this particularly rich strike was not made until the spring of 1880. The mine has produced $400,000. The town of White Oaks was surveyed in May of that year; Las Vegas being the nearest railroad market point. Soon after this first strike the South Homestake, Little Mack, Old Abe, Comstock, Rip Van Winkle and other properties were located, which since became producers of more or less importance.

No mine in White Oaks district is more intimately connected with, or interwoven in the fabric of the history of the camp, than that of the Old Abe.

This mine has been the barometer of life and trade of White Oaks, from its earliest inception as a producer. The Old Abe mine was discovered in the winter of 1879-80; the true vein was not found until November, 1890, when pay ore, in large quantities, was uncovered. The first thirty-five feet in depth proved the value of the ore to average $30.50 per ton. A twenty-stamp mill was erected which began operating in April, 1893; about five years later cyanide was added to treat the tailings and a canvas floor concentrator has recently been installed.

John Y. Hewitt, one of the first owners, has been identified with the Old Abe mine through all its vicissitudes down to the present writing.

The Old Abe vein is a true fissure varying in width from 3 inches to 16 inches; but attaining the extreme width of 22 feet in the chamber known as the "fish pond," which lay between the 7th and 8th levels. The ore is now principally a sulphide; although near the surface it was mainly free milling.

A total depth of 1,375 feet is reached, which is the deepest mine in New Mexico. Practically no water has been encountered and it said to be the deepest dry mine in the world. The total output to January 1, 1904, is $875,000 in gold; but little more than a trace of silver is present in the ore. Virgin gold in gypsum is one of the remarkable occurrences in the Old Abe.

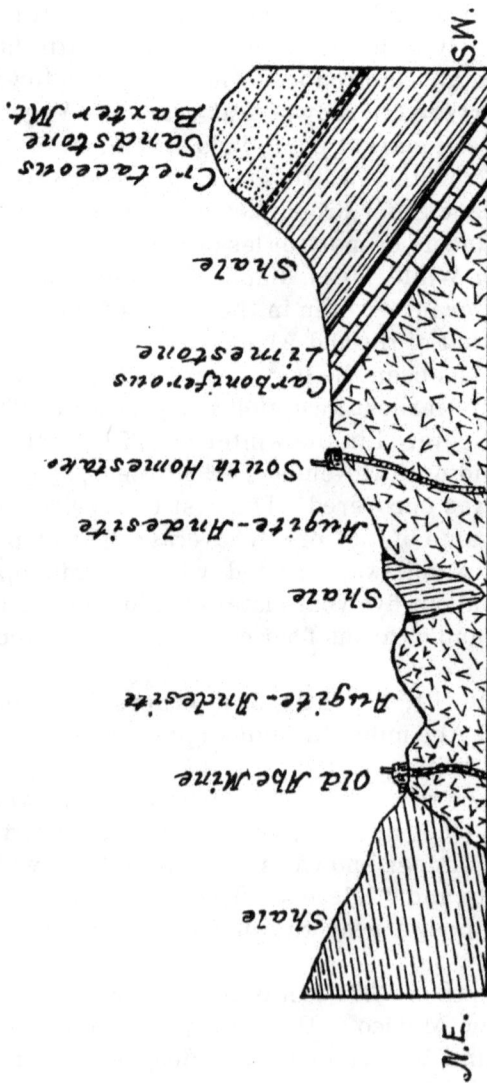

Fig. 31—GEOLOGICAL SECTION OF WHITE OAKS MINING DISTRICT.

The Little Mack has turned out some of the richest ore in the camp; at present the property is tied up in litigation. This property has produced $50,000.

The Lady Godiva was located in 1883 by B. H. Dye. Some beautiful specimens of wire gold have been taken out and the mine has had a considerable production.

The Boston Boy is another property of much merit and and was located in 1893.

Next to the Old Abe in production stands the South Homestake, which is credited with $600,000.

In the Compromise, most beautiful specimens of nugget and wire gold have been found; the property is idle due to litigation.

The Rita, Henry Clay, Little Homestake, Comstock, Rip Van Winkle, Bristol, Thunderer and Little Nell, possess many qualities which go toward making up a mine when properly developed. The total production of White Oaks camp up to January 1, 1904, is $2,860,000.

There are five gold mills in the White Oaks district, one of these, belonging to the North Homestake, is composed of two 5-foot Huntington's; the others aggregate 50 stamps.

The following is a geological section, as sketched by the writer on February 6, 1904, from the shale contact near the Old Abe mine to the top of Baxter mountain.

Successive lava flows to the west of the camp are observed, which would show that much igneous activity had at certain intervals, visited the region. The older or primary eruptions must have genetically influenced and made possible the mineralization which exists in particular and favored zones.

Of the White Oaks range, Baxter mountain seems to embrace the gold bearing area; all of the principal mines of the district lie in a very circumscribed area and on the east slope of the mountain.

To the east is Patos peak, in which vicinity a good quality of coal is found; other indications of coal exist near White Oaks. Considerable deposits of iron ore occur in certain localities to the northeast; Lone mountain is nearly encircled by a good class of iron ore. Splendid building stones from the Cretaceous sand rocks are near by.

Baxter mountain received its name from George Baxter, one of the pioneer prospectors of the district.

Jicarilla District.

The Jicarilla mining district lies to the northeast of White Oaks about ten miles and is about seven miles southeast of Ancho, a station on the El Paso and Northeastern Railway, in Lincoln county.

Ancho peak is the highest in the district; Monument peak, near by, is a close second. The elevation at the postoffice, as determined by the writer, February 4, 1904, by barometric pressure, is 7,475 feet above tide water.

The most prominent land mark in the district is Jack mountain, which occupies the northern end of the range, and is conspicuous for its precipitousness on its north and west faces.

Quite a large deposit of iron ore exists in the vicinity of Jack mountain; the percentage of silica is said to be rather high.

The country rock is an altered granite, due to kaolinization of the contained feldspar.

At one time the Jicarilla mountains were evidently much higher; at least 1,000 feet appears to have disintegrated, exposing the granite core over the major part of the district; in fact, the process of leveling seems to be going on now at as rapid a rate as in the past.

No large veins occur in the region and the seams and seamlets in the basal core run in every conceivable direction; in these seams the gold has its origin. Due to disintegration of this network of small veins would account for the placer gold found so prevalent throughout the district. Usually the gold is fine and does not have the appearance of having been transported far; inasmuch, the particles are angular and in some instances still adhere to the associated quartz vein-stuff. Thus, as the process of weathering progressed the gold, owing to its specific gravity, remained near its source, while the wind and water carried away the constantly forming detritus of the decomposing rocks. The nature of the detritus on the mountain slopes, always sharp and angular, supports the theory of the genesis of the gold, and its accumulation is due to a concentration process resulting from the weathering of the mountains.

Placer mining was conducted in the Jicarillas by the native Mexicans as early as 1850; they would cart the dirt in one direction and bring water from another, and there separate the gold in a sort of wooden bowl called a "batea." Many of the old pits and dumps were still visible in 1880, at the time of the advent of the American prospector.

The American Placer Company inaugurated and operated a large dredge, during the season of 1903, in one of the gulches west of the postoffice; the process proved too expensive to pay.

Numerous claims in both placers and lodes are located and held over the entire district; so far, the principal production has come from the placers. The fineness of the placer gold is about 920.

The following is a list of the principal claims existing in the Jicarillas January 1, 1904, viz:

Mountain Boy group, Geneva group, Iron King group, Admiral Dewey and Belmont, Good Luck, Belle of Memphis, Belle of New Mexico, Summit, Ready Relief, Old Comrade, Little Giant, Eureka, Zulu, Richmond, Central, Revenue, Exit, Annex, Prince Albert, Dark Cloud, Queen Victoria, Jicarilla, Placer company, Knickerbocker, Democrat, Cleveland, Hawkeye, Juana Gulch Placer, January, Jane Anderson and Comery lode.

A belt or zone of copper has recently been found to pass through on both the east and west sides of the district; these are receiving some attention at the present writing.

Red Cloud (Gallinas Mountains) District.

Owing to the great distance, heretofore, from transportation, this mining section has developed but little; the building of the El Paso and Northeastern Railway, has eliminated this obstacle and it is expected to witness a revival of mining in this region in the near future. The district occupies a portion of the extreme western part of Lincoln county and is situated some twenty-five miles east of the Gran Quivira.

The oldest locations in the district are the Red Cloud, Tenderfoot and Deadwood; these claims were opened up a

Fig. 32 — DREDGE OF THE AMERICAN PLACER COMPANY.   Photographed by F. A. Jones, 1904.

number of years ago, and the ore was hauled in ox wagons for a distance of nearly one hundred miles to the Rio Grande smelter at Socorro.

Perhaps, the most promising property is that of the Old Hickory; its ore is principally copper, but carries, also, a good percentage of lead. The Hoosier group has excellent showings, and the same thing may be said of the properties belonging to the American Consolidated Copper Company. The Sunbeam, Buckhorse and Last Call are, also, well known and may be considered promising propositions.

Large iron deposits exist in different parts of the Gallinas; notably, among which may be mentioned the Harris group of iron claims, which appears to be the most valuable property in the district.

Active prospecting and development are, at this writing, July 1, 1904, being prosecuted; the outlook for the district has greatly improved since the first of the year.

Trachytes and a variety of porphyries compose the general rock system. The principal veins are contacts between the lime and porphyry. Plenty of the finest of timber can be had for all mining purposes; water, however, is quite scarce.

CHAPTER XXIII.

COOK'S PEAK DISTRICT.

Rising out of the alluvial plain, twenty miles north of Deming, a monolith of solid granite seems to pierce the blue sky. This solitary sentinel of the unspeakable rocks marks the southern terminus of the rugged Mimbres range, reaching an altitude of 8,300 feet above sea level. Phillip Cook, purported to have been connected with the United States Army about the year 1860, is said to have been the first white man to scale the peak.

Due to this incident, this conspicuous landmark has ever afterward borne the name of Cook's Peak.

The mining district which lies around this pinnacled rock is denominated Cook's Peak, and lies in northern Luna county. This district ranks among the best lead producing mining sections in New Mexico.

The ore is principally a lead carbonate, yet in certain properties galenite occurs, especially on the northwest flank of the mountain at the Jose postoffice. Sand-carbonates also, are found in the mines near the summit at the Cooks postoffice, and elsewhere on this side of the summit.

The axis of the mountain peak has severed the heavy Carboniferous limestones into two parts, one of which reaches from the Graphic mine to a point well up on the mountain pass beyond the Cooks postoffice; and the other is across on the northwest slope, near the foot of the mountain at the Jose postoffice; these two areas embrace the principal mines of the district.

Segregation of the ore has taken place in the usual manner at or in the lime contact, and the chief deposits occur entirely in the limestone cavities.

The principal production has been in the immediate vicinity of Cooks postoffice, at and near the divide, and also at the Graphic mine near the mouth of the canyon.

By far the most important mines in the district are the

Desdemona, Othello and Monte Cristo, belonging to the American Smelting and Refining Company; the Graphic, belonging to the Graphic Mining Company; and the Teel and Poe mines, known as the Summit group.

The first of this classification have produced nearly $2,000,-000; the second or Graphic group is credited with $450,000; and the third or Summit group with $350,000. The remaining properties of the district are entitled to a credit of $200,-000. The total production of the district up to January 1, 1904, is approximately $3,000,000; one-fifth of this amount is silver and the four-fifths is lead.

Among other important producers and prospects may be named the Lead King group, Contention group, Wisconsin and Minnesota, Mocking bird group, Cleveland, Excelsior and Roosevelt.

Over on the Jose side the most important property is that of the Faywood Lead Company, which has recently installed a 50-ton plant, known as the Hooper pneumatic process of concentration.

Should a process of this nature prove successful, it would virtually revolutionize the concentration of ores in a number of New Mexican dry camps.

The White Oak group, Big Galena, Monitor and Bonanza have favorable showings as prospects.

Ed Orr, who is yet living in the district, appears to be the first prospector and discoverer of ore in that region; this was about the year 1876.

No excitement existed in the Cook's Peak region until after the arrival of Taylor and Wheeler, in the year 1880. To these prospectors is due the credit of locating the principal producers of the district.

Their first location was the Montezuma lode; afterward the Graphic, Desdemona and Othello claims were located.

During 1881-2 the outlook of the district as a leading lead producer was practically assured.

J. K. Gooding and Giles O. Pierce purchased the Othello and Desdemona from Taylor and Wheeler in 1882, and afterward sold the property to the Consolidated Kansas City Smelting and Refining Company; this latter company and its holdings were absorbed later by the American Smelting and Refining Company.

George L. Brooks arrived at Cook's Peak in December, 1881, and during the summer of the following year, under an escort of soldiers from Fort Cummings, graded the wagon road up the main canyon from its mouth, to the summit or top of the divide above what is now Cooks postoffice. Indians were very troublesome in those days; Victorio, the Apache Chief, had removed his camp from near the mouth of the canyon, through which the road was graded about two years before. The region of Cook's peak was a veritable stronghold for the Apaches in the later '70s; it was necessary to take every precaution, since it was not known when they would return to their old haunts and ambush the unsuspecting frontiersman.

After completing the wagon road Brooks hauled out 2,700 tons of ore, the first ever taken out of the camp, and delivered a portion of it to the Lake Valley smelter and the remainder was hauled to Florida, the nearest railroad point, for shipment.

Florida District.

Some twelve miles to the southeast of Deming in Luna county are the Florida mountains, which embrace the Florida mining district.

This range of mountains is circumscribed and isolated, being completely surrounded by the great alluvial plain, that extends to the border line of the Mexican Republic, and is broken only in a second place by the conspicuous peaks of the Tres Hermanas, lying about fifteen miles farther to the south and west.

The stratified Carboniferous series flank the range and in a few places it has been lifted bodily upward, being supported by the igneous members beneath. Granite, gneiss, porphyries and limestone are the lithological characteristics of the district.

Segregation of the metallic ores occurs along the contact and in the lime. The ores are sulphides and carbonates of lead, copper and silver.

The district was first prospected in the year 1881. Of the

first prospects, only the Silver Cave lode proved profitable to work; it was discovered by Carroll brothers. Eighteen hundred tons of lead-silver ore were shipped during the time the mine was in operation, which brought $60,000; work stopped the latter part of 1885 and the owners went to Europe. The Silver Cave had been idle since that time, until September, 1903, when Mr. James Carroll returned to Deming and put a force of men to work.

The Silver Cave group consists of the Silver Cave, Pocahontas, Sun Rise and Sun Set, all of which are patented.

Two promising copper claims are owned by John Stenson of Deming; the vein seems to be a fissure in the porphyry near the lime contact. Several other properties, such as the Bear, Tiger, Iron Mass, Lead Carbonate and Roosevelt are well and favorably known throughout the district.

Tres Hermanas District.

The Tres Hermanas district, in Luna county, lies almost due south of Deming twenty-five miles and southwest of the Floridas some fifteen miles. A cluster of three peaks embrace the district from which it took its name.

The ores are principally silver-lead and the general rock formation is almost identical with that found in the Florida mountains.

The Cincinnati mine is the most prominent in the district; it is given the credit of having produced approximately $100,000. The deposit was near the surface; depth failed to prove the permanency of the ore body. Undoubtedly the deposition of ore was chiefly due to descending surface waters.

The Golden Cross and Eagle Mining Company owning the Cincinnati group, control the Judge McComas property at Carrizillo springs which embraces the Yellow Jacket mine and consists of eleven claims.

The Hetty group operated by A. J. Clark and associates of Deming is at present a producing lead-silver property.

Another property of considerable merit is that of the Hancock Mining Company which operates the Hancock group.

Victorio District.

This district is in the western part of Luna county immediately south of the Southern Pacific Railway; Gage is the nearest railroad point.   It received its name from being one of the chief haunts of the Apache Indian Chief Victorio, in the later seventies and early eighties.

The geological formations here are much the same as at Tres Hermanas, Fremont and other mining camps that lie in the southwest central plain; being eruptive types of mountains and flanked by the Carboniferous limestones.   The mineralization occurs along the planes of contact; although, true fissures exist in some instances.   The ores are principally a silver-lead product, and in some cases good values of gold have been found.

Wolframite, or more properly speaking, hubernite is found in this locality.   The St. Louis and Chance mines are the ones which have rendered the Victorio region famous.   Approximately $1,150,000 have been taken out of the district up to January 1, 1904.

Mr. Michael Burke, as lessee of the Chance property, took out nearly $85,000 from that mine within less time than a year.

Stonewall District.

This section of New Mexico lies in southwestern Luna County, at the Hermanas junction of the El Paso and Southwestern Railway.

But little development has ever been done here, although a few of the prospects seem to have some future.   The metallic values are chiefly gold.

CHAPTER XXIV.

COCHITI DISTRICT.

This region lies in the northeastern part of Sandoval county, almost directly west of Santa Fe and on the opposite side of the Rio Grande from that city.

Prospecting was first conducted in this district as early as 1880; Major Edward Beaumont was among the pioneers. At this time nothing favorable was discovered and in the meantime the prospectors were ordered to leave by the Mexican element, who claimed this particular section of the country was a private land grant. Things were dormant until 1889 when an influx of prospectors, notably Joe Eagle, N. R. D'Arcy, Thos. Briggs and others, who made locations.

Eagle and D'Arcy afterward sold to Col. R. W. Woodbury of Denver the Iron King lode. Woodbury, in 1896, erected a milling plant about seven miles below the present site of Bland, and now known as the Woodbury mill.

In May, 1893, the Ellen L., Mammoth and Washington locations were made by a Mr. Pilkey for himself, Henry Lockhart and Ben Johnson.

In the fall of the same year a boom swept over the district and the entire country was covered with locations.

The Crown Point group, consisting of the Crown Point, Giant and Bull-of-the-Woods, was located by J. D. May for himself, Henry Lockhart, Ben Johnson and B. F. Bruce. The Posey, Black Girl, Little Casino, Allerton, Union and others were located by George Hopkins and partners.

In December, 1893, at the end of the month, the Lone Star group of three claims, was located by Chester Greenwood, Norman Blotcher and Henry Woods for themselves and Thos. Lothian of Denver.

The noted Albemarle group, which made the camp famous, consists of the Albemarle, Ontario, Pamlico and Huron, was located by the latter parties about the first of January in 1894.

The Albemarle group passed into the hands of the Cochiti

Gold Mining Company, a Boston syndicate, which developed
the mines extensively and in 1899 late in the season completed
a large plant of 300 tons capacity; the electrical power being
supplied and transmitted a distance of 35 miles from the coal
mines at Madrid to the mines.

The ore was reduced by dry crushing rolls to pass through
a 24-inch mesh and the slimes were removed by exhaust fans
—cyaniding was the process of gold extraction.

The plant operated for two and one-half years, producing
$667,500 in gold and silver, closing down indefinitely in the
spring of 1902; since which time the property has been in the
hands of a receiver and the plant partly dismantled.

The cause of the failure is attributed to the gradual de-
crease in the values of the ore as depth was reached; near
the surface the ore average $6.50 per ton, at the extreme
depth the ore did not exceed $3.50 per ton.

The ore body in the Albermarle was quite considerable,
varying in width from five upward to nearly sixty feet. The
main shaft is about 700 feet on an incline of approximately
45°.

From a geological standpoint the Cochiti district is extremely interesting. A central core of augite-andesite is
encompassed by a vast field of tufa of a trachytic and brecciated nature. This enclosing field is destitute of any mineralization; evidently, the andesite is the ore-bearing member of
the eruptive series of rocks. From an examination of the
nature of the ore bodies in the district it is inferred that
such mineralization is due to circulating descending hot
waters, which in their downward course have generally en-
riched the veins nearer the surface more than at greater
depth. The segregation of such mineral values must have
taken place during and immediately after the trachitic tufa
outbreak.

In Peralta canyon a curious example occurs in the presence of uranium and vanadium oxide filling the interstices of
a silicified volcanic breccia found on the property of the
Peralta Gold Mining and Milling Company.

The general character of the ore throughout the district is
rather low grade, indurated, flinty, bluish-white quartz.
Much of it is impregnated with small particles of pyrites of

iron; this latter feature is more marked at depth where the ore becomes more refractory. The principal value is in gold; the ratio between the gold and silver being about two to one.

The Lone Star is a prominent claim, having shipped a quantity of ore averaging about $40 to the ton.

The Washington joins the Lone Star on the south and is extensively developed; it is credited with some production.

About one-half mile north of the Lone Star is the Crown Point, which is thought to be a spur from the main lode. The Laura S, Tip Top, Iron King and a number of others have been recognized at times as producers. The Iron King, heretofore mentioned, has the distinction of being the first claim discovered and located in the district. In Pino canyon are the Mammoth, Ellen L. Hopewell, Good Hope, Iron Queen, Black Girl and a number of others having more or less development.

Above the Bland postoffice on the north side of the canyon is the Bull-of-the-Woods, which vein has been opened on the Little Casino. Near the Washington are the Fraction, Short Order and Iowa No. 2.

Passing further south are the Monster, Corona, No Name and Little Mollie; the latter claim has three distinct lodes. In Colla canyon are several locations which if properly developed might prove productive.

Still further to the south in Peralta canyon are numerous locations; in this canyon is where the uranium and vanadium oxide occurs as already mentioned. The Tom Boy, Santa Fe group, Hanison, Famous, Old Dutch, Little Betsey, Ivanhoe, Aunt Betsey, Acme, Calumet, Union, Del Fino, Morning Star, Shannon, De Wit, Strip No. 2, Belle lode, Navajo, Golden Cochiti Tunnel Company, Sister C., Cross Keys, Puzzle group, Midnight group, Wilson and Sheridan group are some of the better known locations.

Nacimiento District.

This district embraces the Nacimiento or Jemez mountains and lies in Sandoval county on the west side of the Rio Grande.

To the southeast, the country is prominent for its mineral

springs and sulphur deposits. This latter section is, also, equally noted for the many prehistoric ruins found near the Jemez and Sulphur springs and elsewhere in that vicinity.

Copper and coal seem to be the minerals of the greatest economic importance. The copper is found chiefly in sandstone on the west slope of the Nacimiento mountains.

The Nacimiento range is very uniform in height and forms a single ridge and is unbroken by any sharp peaks; its greatest elevation is 10,045 feet above tide water. Its extreme length is about fifty miles north and south and the ends gradually pass into and blends with the great level plateau which lies mainly to the west. The backbone or axis of the mountain ridge is a red granite, which forms the principal and elevated portion of the rock-mass.

The geological sequence from this granite core, is the Carboniferous, Permo-Carboniferous or Permian, Jura-Triassic and the Cretaceous; all of which are exposed and readily recognized.

The copper appears to have come from the saliferous beds of the Permian; since it occurs in the conglomerates and sandstones which immediately overlie that series. Precipitation of the cupric solutions seems to have been due to the presence of carbonaceous matter, rather than to the association of iron. In evidence of this may be seen the impress of leaves, reeds and portions of trees which have been completely transformed into some forms of copper ore, preserving permanently their original outlines. Some teeth of extinct saurians have been taken out of these beds, which are classed as odontolite, by reason of their cupriferous impregnations.

The principal copper properties of this section of the Nacimiento mountains are owned by the Jura-Trias Copper Company. This company has done extensive development and improvements and own about twenty-five claims, or 1,000 acres.

This mineralized area is approximately ten miles long by a mile wide. The dip of the formation averages 45° toward the west, and the mineralized zone or stratum is, perhaps, 50 feet thick.

Of course, only certain portions of this zone are capable of being profitably worked.

Some evidence exists of early Spanish mining in the district; the proof, however, is not conclusive.

To the west a short distance is a cropping of coal; so far as known the coal is not coking.

Dr. J. S. Newberry, who accompanied the Macomb Expedition in 1859, was the first to examine into this part of the country from a scientific standpoint.

A man by the name of Moore, in 1881, did the first modern prospecting in the district; these original locations are now embraced in the holdings of the Jura Trias Company.

San Miguel District.

Some ten or twelve miles to the south of the Jura-Trias Copper Company's property in the same range, is the San Miguel district. Only three locations exist in the district; known as the San Miguel group.

This property was discovered and located in 1882, by Pat Carrol who died on January 4, 1904.

The occurrence of the copper here is identically the same as that described under the Nacimiento district; in fact, it is the same copper bearing formation. Large quantities of low grade malachite exist in the sandstone, accompanied by kidneys of high grade copper glance.

For a great many years the San Miguel property has been owned by J. J. Gorman of Albuquerque, who died about the middle of April, 1904.

CHAPTER XXV.

MINING DISTRICTS OF THE SANDIA AND MAN-ZANO MOUNTAINS.

These two mountain ranges belong to that class of titled orogenic blocks that are typical in New Mexican topography.

In all respects they are identical with the Sierra Oscura and San Andreas ranges, of which mention has already been made.

While these two mountain ranges are more or less mineralized, producing paying mines appear to be foreign to them up to the present time. The Sandias, however, produce plenty of good building stone and afford most of the lime used in Albuquerque. Beds of good refractory clay and graphite abound in the region.

Placitas District.

On the extreme northern end slope of the Sandia range, in Sandoval county, is the Placitas district. Not until recently has this district been active; although it has been said to possess some merit for a number of years. Copper appears to be the chief metal, which is associated with lead, silver and gold.

The horizon which seems to carry the best values is on the contact of the Carboniferous limestones; many of these veins are "blanket" in form.

Some of the principal groups and claims are the W. J. Bryan, Nineteen-hundred group, Balcomb claims, Shamrock group, Bibo, Iron Cap, Montezuma group, Yellow Jack and Valley View group.

Sandia District.

Lying toward the east from the Placitas district in the direction of South mountain and on the northeast slope of the Sandia range, is situated the Sandia mining district.

Auriferous beds of cement gravel occur along the foot of the northern slope of the Sandias; these gravel beds are now receiving some attention, and may finally be worked at a profit.

The southern and eastern parts of the district are conspicuous by the red character of the formation, which is recognized as the Permian or "red beds." In these rocks which are tilted to the south and west at an angle of 35°, cupriferous zones or bands exist and are very similar to those described in the Sierra Oscura, further to the south; the organic matter having been replaced by copper compounds which retain the imprint of the fossil plants. This extraordinary condition of mineralization in the red sandstones is very marked, as observed in an examination made by the writer on the Texas group of claims, May 18, 1904, for the Pennsylvania Copper Company, in the Sandia district.

The Gold Ring and Maceo groups lying in another part of the district, apparently have a different class of vein from that of the Texas, appearing to be of the fissure type. The former group is entirely gold bearing, the ore being a free milling quartz, while the latter group carries a sulphide ore containing gold, silver, copper and lead.

Tijeras Canyon District.

It was was from the canyon which cuts through the heavy limestone plate, into the very core of the granite basement in an east and west line, dividing the Sandia mountain range into two parts, that this district took its name.

Tijeras canyon has become noted as a resort for Albuquerque people, the chief attraction being the famous Whitcomb Spring.

In this mining district prospecting has been carried on from time to time covering a period of fifty years; its proximity to Albuquerque makes the region a favored one for the "tender foot."

In certain localities along the lime and porphyry contact the sulphides of iron and copper have segregated, carrying some gold and silver values. No prominent producing mines have ever been opened up in the district. There are, however, some prospects which appear most promising. The Carnuel group, Silver Dollar, Long View and Magnolia No. 1, are worthy of being mentioned. The Silver Dollar is owned

by G. Oxendine and son (darkies), which they have developed
in a very creditable manner.

Coyote and Star (Hell Canyon) Districts.

These two district lie south of Tijeras canyon and are
approached from the southwest.

In Coyote canyon is the popular Coyote Springs water:
these springs are located at the west end of the large quartzite
dike that caused the rift in the Sandia range, and near the
eastern end of which is the Whitcomb Spring.

Some prospecting was done in this canyon by Americans
in the early days, and it is also thought that the early
Spanish explorers prospected for gold near the Chavez Spring,
since some vestiges of old pits yet remain.

Placer gold is said to exist in the gulch gravels of Hell
canyon and the Reliance Gold Mining Company was organized
on the strength of this supposition and proposes to erect a
plant to treat the auriferous gravels.

The Malagras and Golden groups are also well known in
the district.

Manzano Mountain District.

Nothing of any real value in a mining enterprise has ever
been conducted in these mountains, and much the same conditions exist here as in the Sandias lying to the north.

Within the past two years since 1902, quite a number of
copper prospects have been located in the vicinity, and to the
north of Abo Pass.

The copper occurs in a gray sandstone, disseminated as
carbonate through the rock, and is very similar to the copper
bearing sandstone near Las Vegas. Some very high grade
ore is found, though on the whole the average is low, not
exceeding 4 per cent in metallic copper.

The most prominent property here is a group of nine locations owned by the Belen Carricito Development Company.

CHAPTER XXVI.

SILVER HILL (Jicarilla) DISTRICT.

Far out in the broad expanse of desert in southwest Otero county, may be seen a low isolated group of mountains rising above the plain, having the appearance of an island.

This desert isle is surrounded by a sea of drifting sands and lies near the center of the plain, between the Organ and the south end of the San Andreas mountains on the west, the Sacramento range on the east and the waste of the mysterious "white sands" immediately to the north. This dot in the desert, though having the dimensions of twelve miles in length by four miles in breadth, embraces the Silver Hill (Jarilla) mining district. The Jarilla mountains were, perhaps, never very high, yet they have been greatly reduced by erosion from their original size. Due to this erosion are the placer deposits at the southeast part of the district.

The oldest rocks are, perhaps, pre-Cambrian; granite occupying the northern portion of the range. Flanking and dipping away from this granite core are the sedimentary series; yet, throughout the district considerable areas of these sedimentaries, have been elevated and repose on the subjacent granites. The general elevation of the district must have taken place at the close of the Cretaceous or beginning of the Tertiary.

Andesite and trachyte flows with intrusives of granite-porphyry and diorite have played an important role in the dislocation of the rock system and in the immediate subsequent segregation of ore. Some instances are found in which the porphyry has entirely surrounded the Carboniferous limestone; this phenomenon is found on the property of the Three Bears Mining Company.

Extensive faulting seems to have taken place after the deposition of the ores. The strike of the fault-planes are two fold; north-south and east-west.

It is observed, on the property of the Three Bears, the east and west faults, dislocate the north and south veins.

There are two systems of fissure veins in the district, cutting alike all formations. The older system has a strike N. 15° W. and is badly faulted; the younger system has a strike N. 70° W. and is undisturbed. The dip is about 80° toward the east, in the main system of veins of the Three Bears.

Contact metamorphism has greatly influenced the adjacent Carboniferous limestone, resulting in the formation of massive garnet at numerous places. It would seem that the porphyritic intrusives are intimately connected with the origin of the ore. These mineralized dikes, through the seductive influence of the Carboniferous rocks, were induced to impart with their metallic compounds along the planes of contact. The principal commercial minerals of the district are iron, gold and copper, the latter metal exists under most every form of copper ore known. This district is, also, noted for its turquoise, which bears evidence of prehistoric working.

Prospecting was first conducted in the Jarilla mountains by S. M. Perkins in 1879; the district, however, did not come into prominence until nearly twenty years later, when Amos J. DeMueles made it famous through its turquoise discoveries.

No turquoise mining was done in the Jarilla mountains during 1904; the mines being substantially closed down since the killing of DeMueles by Jacobo Flores, a Mexican boy, which occurred at Jarilla in 1898. The Three Bears Mining Company has by far the best developed property in the camp. A depth of more than 200 feet has been attained, with drifts at each hundred-foot level. The gangue matter is chiefly calc-spar above and at the 100-foot level; but, at and below the 200-foot level this spar has given place to soft masses of honeycombed quartz, that occasionally carry fair values in gold and copper. This quartz has evidently been leached to a large degree of its values, and the metallics, no doubt, have been redeposited at some greater depth. This lode of the Three Bears can be traced nearly two miles further north. So far as known, only one other north and south vein exists in the camp and that is the Nannie Baird lode, on the east side of the range. Quite a large production of gold is accredited to this property.

Of the younger veins the Little Annie and Garnet are examples. The former is in porphyry and an ore shoot quite rich in gold has been developed. The Garnet lies two and one-half miles north of the Three Bears.

Several deposits of iron of good commercial value occur in different parts of the district.

Fig. 33 –DRY WASHER OF THE ELECTRIC MINING AND MILLING COMPANY. Photographed by F. A. Jones, 1904.

The iron carries some gold; about 1,500 tons were shipped from the Iron Queen lode in 1903, to the El Paso smelter. In some places, especially at the Electric Placer Company's property, part of the limestone has been replaced by the iron.

Placer mining is being carried on to a certain degree of success by the Electric Mining and Milling Company; the apparatus employed is a hand dry-washer. In June, 1904, the largest gold nugget ever found in the district was taken out by this company; it weighed 6½ ounces and was valued at $123.00.

Near bed rock or hard-pan the values per cubic yard are $4.00. The fineness of the gold is 940.

Arrangements are being made to work these placer fields on a much more extensive scale; provided water can be had in the well now being drilled. The water question is a serious one; since none is had excepting that hauled into camp by the E. P. & N. E. Railway. The south end of the district is distant from Jarilla Junction (a point on the main line of the E. P. & N. E. R. R.), about three miles.

Some of the principal groups and lodes of the district beside those already mentioned are the following:

The Alabama group, Last Chance group, Monte Carlo group, Alice group, St. Louis Copper Company's mines, By Chance group, Altamont Mining Company's property, Penarilla Mining Company's group, the Lucky, Lincoln, "7 come 11" and Maggie lodes and the North End and Red Hill groups.

CHAPTER XXVII.

SELITRE, MINERAL HILL, SAN PABLO, SAN MI-GUEL AND TECOLOTE (Las Vegas) DISTRICTS.

During the fall of 1900 attention was first drawn to the immense ledges of sandstone that were impregnated with copper, covering large areas to the west and southwest of Las Vegas, in San Miguel county.

This region embraces the mining localities of Selitre, Mineral Hill, San Pablo, San Miguel and Tecolote, none of which are well defined districts.

While the existence of copper was known throughout this section for many years, no definite steps were taken toward its exploitation and development until 1900.

The sandstone in which the copper is present resembles in some particulars that found along the east side of the Sierra Oscura at Estey City and various points further to the south. Similar outcrops, though less extensive, exist in the Chupaderas.

It seems that the cupriferous impregnations have been leached from the Permian beds and redeposited in a particular layer of subsequent sandstone, presumably Triassic. The deposits are usually "blanket" in form and as a rule of very low grade; averaging less than three per cent of metallic copper. The type of the ore thus found, is principally malachite; although, occasional pockets of high grade glance are encountered.

Leaching appears to be the only process by which the ores can be successfully treated; two experimental plants have quite recently been making trial runs, which seem to offer a fair degree of success.

Evidently, the time is near at hand when almost every ton of this ore will be subjected to treatment.

A small advance above the present price of the red metal will likely see Las Vegas, within the near future, assume all the dignities of a typical copper camp.

Prominent in the Selitre district are the Selitre, Copper Queen, Last Chance and Santa Isabelle groups; in the San Pablo district, the Overflow, Gonzales and Santa Rita groups constitute the principal properties; and in the San Miguel and Mineral Hill districts may be mentioned the Copper King, Santa Maria and Mystic groups.

Prospecting in this region, until recently, has been gravely hampered and handicapped; the chief hindrance being due to unconfirmed land grants, which barred absolute title and possession.   Since the matter of title has lately been finally adjusted by the courts, prospectors are now more active in their work.

Rociada District.

North of Las Vegas, in San Miguel county, near the Mora county line, in the vicinity of the village of Rociada, a new mining field is being developed.

The Rociada district had its birth in June, 1900.   Since that time much prospecting has been done; copper associated with iron seems to be the predominating feature of the camp.

The formation is granite with associated porphyries. These are traversed by mineralized dikes having a general north and south strike.   The mineralized zone seems to be about five miles in width and twelve miles in length.   Small values in gold and silver occur associated with the copper-iron ores.

Considerable development has been done on the Rising Sun group, owned by Governor O. A. Hadley and others.   Parlleling the vein of the Rising Sun is the Azure lode, which is also being developed.   The Joe and Jennie group, the Lone Star and numerous locations have been made and considerable work done. The district has never produced, the grade of the ore being insufficient to stand the long wagon haul to the railroad.

The occurrence of lead and zinc has been noticed in a small way.

Coyote District.

Still farther to the north in the vicinity of the Guadalupita postoffice, Mora county, is some copper property which has been developed to a considerable extent.   The ore occurs in a black and bluish shale.   Malachite predominates, though high grade copper glance occurs sparingly, associated with the carbonate compound; the grade of the ore is low.

This property is known as the Overton group.

Miscellaneous and New Districts.

In various parts of New Mexico, in no recognized districts or localities that are difficult to describe, prospecting is being carried on with possibilities of success.

Near the Willis postoffice on the upper Pecos the Edna group of mines are being developed; the ore is a sulphide and carries values in copper, gold and silver. In the Bernall district is the Maggie group, carrying malachite.

In the Cooper district, in the upper Pecos region, are the Barthe group and the Willow Creek group; the latter propperty belongs to the Pecos River Mining Company. Near Glorieta and north toward the Rio Pecos, R. A. Bradley, the hermit miner, has done extensive developement on several properties of gold, silver, copper and lead.

Within three miles of Santa Fe, are found mineral indications that will doubtless receive attention some time. This latter region abounds in copper, gold, silver, coal and iron. The Sunset group of claims lie about three miles northwest of Santa Fe and are being developed under the supervision of A. R. Gibson, the present mayor of that city. Near Monument Rock, about nine miles east of Santa Fe, large ledges of low grade gold ore are said to exist; considerable development has been done there. On Indian creek is the Annie Jones group, which is very favorably located, and seems to have a future.

Along the "scenic route" leading from Santa Fe to Las Vegas a number of lode mines are being developed, especially in Dalton canyon. In the vicinity of the Gravel Cabin, and below that point there are some fairly good sized veins of coal being opened up.

The whole of the country lying to the northeast of Santa Fe, covered by the Pecos Forest Reserve, is known to be mineralized, but is generally passed by, since the average prospector prefers to avoid any complications which might arise from being on a government reservation.

District in Zuni Mountains.

In the Zuñi range of mountains in northwestern Valencia county a number of camps and small mining districts exist. None of these districts, perhaps, has ever been fully or properly organized and all of them are referred to the Zuñi mountains for want of better description.

Copperton is the most prominent of these districts and is the best developed section; it lies near the center of the mineralized region.

Since the country is new, our knowledge of the district is necessarily limited.

It appears that there are three principal belts or places in which the ore deposits are the most prominent. In township 10 N. range 11 W. of the New Mexico principal meridian, in which township the Montezuma Mining Company is located; township 11 N. range 12 W. embraces the Conductors company and others and which also includes Copper Hill; and township 10 N. range 12 W. contains Copperton postoffice and the Copperton Mining Company.

North of Copperton the Gold Lode Tunnel Company are developing an iron dike which lies in limestone and porphyry. The Fourth of July location has a shaft down 100 feet; it belongs to the Montezuma Company. The Compromise lode is a promising prospect.

Considerable activity is manifested in the Zuni country and it would appear that not a great while will elapse until the worth of the region is known.

Guadalupe Mountains.

This low isolated range extends in a southeasterly direction and lies partly in the counties of Chaves, Otero and Eddy.

The section has been but meagerly prospected, though the indications are good for copper and gold. The old Nymeyer copper mine is the best known; the Standard Old Company of Carlsbad recently purchased this property and contemplate extensive development in the near future. Numerous locations along various parts of the range have been made during the past year and good results are expected by those who know the country best.

CHAPTER XXVIII.

IRON.

Behold me! Though rusty and covered with moss I am king of the metals. My ribs of steel girdle the globe. Vitalized by the breath of steam and electricity the commerce of the world rides on my back.

Iron is the most abundant and widely distributed of the useful metals.

The increased use of iron, indicates periods of greatest prosperity.

Iron forms compounds with several of the elements, chiefly sulphur and oxygen. The principal ores of iron are pyrrhotite, pyrite, siderite, magnetite, hematite and limonite. The two first are sulphides, the third a carbonate, the fourth and fifth oxides and the last a hydrous oxide; the three last are the chief ores from which comes the world's supply of iron.

Very little has yet been done to exploit the iron fields of New Mexico; hence, our knowledge as to their extent and character is necessarily superficial. Only one deposit is now being worked; this is at Fierro, Grant county, operated by the Colorado Fuel and Iron Company. The production for the year 1903 was 137,269 tons.

The principal iron field seems to lie in eastern Socorro and western Lincoln counties; although, deposits of more or less magnitude are distributed throughout the Territory.

Generally, the iron is of fair quality and is suitable for the making of good steel. On the whole, a greater quantity of phosphorus is contained in the western iron ores, than is found in the deposits about the great lakes. This objection is overcome to a certain degree by the low percentage of silica.

The iron ores at Fierro run very low in phosphorus; but, much of the iron is rendered unfit for steel and other purposes, on account of the presence of copper and a high percentage of silica. As a general rule all of the large deposits of iron ore in New Mexico lie on the contact between limestone and porphyry. There is scarcely an exception to this rule. Occasionally, deposits are met with where, seemingly, no lime had ever occurred in juxtaposition with the iron; but a careful examination will either reveal this condition existing, or it will show that the limestone has been removed by erosion; since the lime always forms the top covering or encases the iron. This top covering or hanging wall is in nearly all cases a very hard blue limestone of variable thickness and generally destitute of fossils, on account of metamorphism. The iron in most every instance, rests on an intrusive porphyritic sheet or dike. This condition being so general, the first supposition by the writer was, that the genetic existence of the iron was due in some manner to the dike or intrusive sheet.

A study of the origin of most of the New Mexico deposits of iron ore can best be made at the Jones iron group, in the Jones mining district, eastern Socorro county, which seems to be typical, in a general way, throughout the Territory. These deposits are, perhaps, the most extensive in the Territory; extending as they do along a great monzonite dike which runs a little south of east, for a distance of ten miles, cutting at right angles across and through the north end of the Sierra Oscura. Nearly south of this main deposit is a much smaller one occurring on a similar parallel monzonite dike; the geological conditions of the two are identical. It is more than probable that the gnesis of these deposits is due to the leaching of the iron values from the surrounding field of Carboniferous limestone. The rounded and peculiar shape of many of these iron aggregations would indicate that they had previously filled a cavity; in other instances the iron appears to have replaced stratas of limestone of uniform thickness. Moreover, the accumulation of the deposits has a bedded or stratified appearance which would indicate the presence of water in the cavities at the time the iron was precipitated and laid down.

Further, the peculiarities in the roughness of the limestone and the many caverns and depressions or sinks indicate the circulation of much water at one time. Underlying some of the limestone covering or capping and in the horizontal bedded seams, all through the district, small isolated patches of iron stained deposits exist, which became stranded during the time the leaching was in progress. At the time of this geological period, about the close of the Cretaceous, heavy vegetation evidently existed on the lands projecting above the seas. Decaying organic matter was carried into the earth and the limestone rocks by circulating meteoric waters; the organic matter coming into contact with the particles of iron oxide which were disseminated through the rocks, were reduced to the ferrous state and thus passed into solution. These solutions finally reached the ever enlarging limestone cavities and were there precipitated on exposure to the air, as hydrated iron (limonite). It was in this manner that these beds of iron were laid down.

The deep carving, cutting to the very base of the Carboniferous series to the porphyries below, naturally created a line of weakness or least resistance which favored the elevatory movement of the huge monzonite dike which gradually shouldered itself upward supporting the great masses of limestone and iron on its back.

This disturbance necessarily checked the overlying rocks and was accompanied by escaping sulphuric acid gas which in turn attacked the limestone resulting in the formation of the associated beds of gypsum. The limestone mold which originally acted as a support and determined the size and shape of the deposit was thus, eventually carried away and left some of the masses of iron ore stranded on top of the dike, whilst other aggregations slipped down its sides and now rest at or near the bottom of the porphyritic intrusive. Sometimes the iron is found on one side and then on the other side of the dike as well as on the top as before designated. Slicken-sides are frequently observed on the iron which bear out the theory that deposition of the ore took place prior to the elevation of the dike. Owing to the contact of this igneous dike with the lime and iron, some metamorphism resulted in a somewhat mild degree in both the iron and the lime.

The iron was dehydrated and most all of the marcasite which was formed with the limestone at that time was desulphurized. Immediately along the contact of the blue limestone most of the fossils were obliterated.

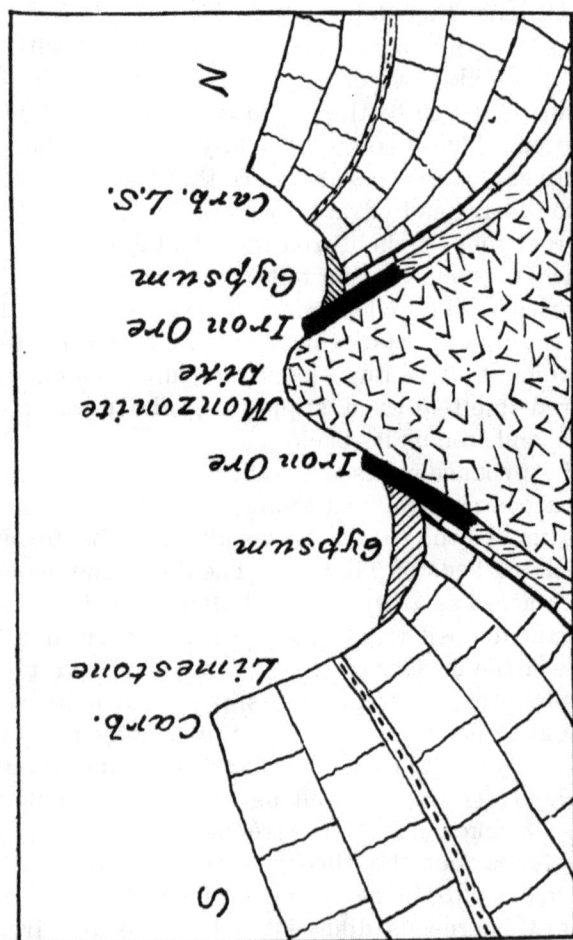

FIG. 34—CROSS-SECTION OF THE MONZONITE DIKE; AT THE JONES IRON DEPOSITS.

Such an exegesis genetically concerning the iron deposits of the Jones district, will virtually apply to nearly all of the other deposits in New Mexico.

To the north some twenty five miles are the Blackington and Harris deposits, discovered by W. T. Harris in April, 1903.

Fig. 35—MASSIVE IRON CROPPINGS, JONES GROUP.   Photographed by F. A. Jones, 1902.
(P. C. (Pete) Bell, Prospector of the Early '80s, in Center of Picture.)

The mode of occurrence of iron at this point is identical with that in the Jones district.   Here the iron is found at intervals on either side of a similar monzonite, dike, though the deposits are not nearly so large.   This locality is in the Chupadera mountains, Socorro county, and westerly from the prehistoric ruins of the Gran Quivira, about twenty miles.   East of the

Blackington and Harris deposits and southeasterly from the
Gran Quivira about fifteen miles, in the Gallinas mountains,
is another large deposit, the extent of which is not known.
This propety is known as the Harris group, discovered by W.
T. Harris and Lon Jenkins in March, 1903, and is in the Red
Cloud or Gallinas mountains mining district, in western Lin-
coln county.

Southeasterly from the Gallinas mountains about ten miles,
and north of Tecolote station three miles, on the El Paso and
Northeastern Railway, is another deposit of a very fine grade
of iron ore.   This is known as the Rock Island group of iron
mines, and was discovered by P. C. Bell in July, 1903.   The
property is very favorably located as the E. P. & N. E. Railway
runs through the same.

Southeast of Tecolote some fifteen miles distant is the
Jicarilla mining district.   A number of good deposits of iron
are to be found in the vicinity of Jack mountain, Lincoln
county.   The iron in this district is of fair quality, being
somewhat more silicious than the ore bodies above described.   White Oaks seems to have considerable iron in its
immediate vicinity.   These deposits are very similar in
character to those already described, and have been exploited
but little.

In the vicinity of the upper Rio Pecos and Las Vegas, good
sized bodies are reported, but lack verification.

A. J. Frank of Algodones and associates, have some iron
locations in the north end of the Sandia mountains in the
Placitas district; this deposit is of the bog-iron character. In
the iron mountain at Elizabethtown quite a deposit of hematite
occurs.

The Kennedy mines at Glorieta have been developed considerably, and the ore at one time was extensively mined and
shipped.   The nature of this deposit is somewhat different
from the other deposits; although its genesis is virtually the
same.

Some properties in the vicinity of San Pedro, known as the
Perry deposits have been exploited somewhat.   The ore carries a little gold and is said to be very silicious.

In the Cuchillo range of mountains, south of Chloride, a
very large body (mountain) of ore is said to exist; this is

known as the Iron Mountain group. Owing to its great distance from transportation the material cannot be utilized at the present time. From the Iron Queen lode in the Jarilla mountains 3,000 tons of ore were shipped during 1903; other properties here seem favorable.

The most prominent iron mines in New Mexico are situated in the Hanover gulch at Fierro, in the Central district. This property has long been known to exist; its early history is contemporaneous with the celebrated Santa Rita and Hanover copper mines, when its magnetic properties attracted attention, in the early part of the 19th century.

As mentioned near the beginning of this chapter the property is being worked by the Colorado Fuel & Iron Company. The ore is shipped to the steel plant at Pueblo, Colorado. This iron rests on porphyry and is in contact with limestone. The iron evidently has been leached from the adjacent rocks through the agencies of organic matter and meteoric waters. Considerable quantities of maganese are found at times in various parts of the deposit.

In the Canyoncito district, Socorro county, just across the Rio Grande northeast of San Acacia, numerous croppings of iron ore exist, apparently in the granite; its character is very silicious.

South of Hachita toward the Big Hatchet mountains one or two fairly large deposits of iron ore have been reported; these finds have not yet been verified.

A large deposit of manganiferous iron ore lies west and adjoins the town of Silver City. Considerable development has been done on the property and much of the ore was used as flux in the Silver City Reduction Plant, when the same was in operation a short time ago. This ore, generally, carries a high percentage of silica and quite a percentage of manganese.

On the whole, New Mexico appears from a cursory standpoint, to be well endowed with great quantities of iron which will shortly command the attention of investors. In fact, the true mineral wealth of the territory unquestionable lies in her extensive deposits of coal and iron.

The following is an analysis of an average sample of the

Fig. 36.—FIERRO IRON MINE. Photographed by T. H. O'Brien, 1902.

best grade of iron ore from the Jones mining district, as made in 1902, by the Seamon Assay Company of El Paso.

Silica	2.00%
Iron	66.40
Lime	.30
Magnesia	1.25
Sulphur	.07
Manganese	trace
Alumina	.50
Phosphorus	.09
Moisture	.20
After ignition (gain)	.77

This iron is magnetic to a certain extent, in a number of places.

Analyses of two samples of iron from the Blackington-Harris deposits, which lie to the north of the Jones district, are here given:

(Elston F. Jones, analyst, May 22, 1903, University of New Mexico).

Constituents.	No. 1.	No. 2.
Iron	58.7%	64.1%
Silica	13.6	7.9
Lime	3.0	1.2
Manganese	trace	trace
Phosphorus	trace	trace
Sulphur	trace	trace

CHAPTER XXIX.

COAL.

A lump of coal may be regarded as a receptacle in which a definite amount of energy from the sun has been effectually bottled up, throughout an indefinite period of time.

In other words, a piece of coal may with propriety, be termed "a parcel of stored sunbeams." This entrapped sunlight represents undissipated energy of the geological past. The enormous energy pent up in the coal fields of the world is beyond comprehension.

By the magic touch of the chemist's wand, this black product was forced to restore the beautiful colors of the aniline dyes to the "temple of the sun" and the "bow of promise" from whence they were taken millions of years ago.

Carbon is among the most subtle and wonderful of all the elements in its selective and polygamous habits; its combinations with oxygen and hydrogen seem to be practically unlimited. Its union with these two elements forms coal, which product belongs to the hydro-carbon series. It appears that the decomposition of vegetation under certain conditions, in the earlier geological periods, is the origin of the coal deposits as we now find them.

The elementary composition of all these substances is practically the same; and there seems to be a gradual passage of plant tissues through all the intermediate stages leading up to and terminating in graphite.

The arrangement of this combination of the hydro-carbons is best shown in the following table:

Substance	Carbon	Hydrogen	Oxygen
Wood.........	4.75	12.18	83.07
Peat.............................	34.48	9.85	55.67
Lignite...................... ...	49.21	8.37	42.42
Coal........	72.65	6.12	21.23
Anthracite....	95.42	2.84	1.74
Graphite.....	100.00	....	....

It is claimed that coal was first discovered in the United States, in 1679, at a point near the present site of Ottawa, Illinois.

The discovery of coal in New Mexico is not attributed to any particular locality or person. The first mention of its existence in the territory was by Governor Chacon, in 1803, who says, "that copper is abundant and apparently rich, though no mines* are worked, though there is much coal of good quality."

The first vein of coal opened in New Mexico was in the Carthage field, at what is known as the "Government mine." General Estanislado Montoya mined coal from this place in 1863, and hauled the product to Fort Craig to supply the needs of that military post during the Civil War. It was from this connection that the mine received its name.

General Montoya claimed, at that time, the whole of the Carthage field, as being a part of his private land grant, which later was declared by the Court of Private Land Claims as part of the public domain.

The second coal mined was at Madrid in the Cerrillos field, in 1869. Work was done here in two localities by the New Mexico Mining Company. At the first of these places the development consisted in two openings from which 280 tons were mined, which the company used for steam purposes in their stamp-mill, at the Old Placers, near by. The other point of work was a short distance to the southwest from the first openings; 100 tons were piled on the dump ready for use. In both localities the work was done on one of the anthracite veins. These observations were made by R. W. Raymond in 1870; and both were on the anthracite vein.

Some of this anthracite coal was tested at Santa Fe, by Mr. Brucker in his assaying furnace, at that time. He states that he was able to obtain a white heat in a very short time and that its lasting qualities were about three times as long as that produced by an equal weight of charcoal.

Coal was known to exist in 1870, at several other places: at a point about ten miles south of the anthracite deposits at Madrid; near Galisteo creek; on the Pueblo Indian reserva-

---

*This verifies the statement that the Santa Rita mine had not begun operations prior to 1803.

tion near Taos at the foot of the Pueblo mountains; on the
Vermejo, Raton mountains, near Maxwell's—vein six feet
thick; on the Purgatoire river, Las Vegas; at the Rio Puerco;
in the San Mateo mountains; and several places west of Fort
Wingate. Approximately 400 tons of coal were produced in
New Mexico in 1870.

Many of the coal veins in the territory attain a thickness
of six feet. The thickest deposits are supposed to occur in
San Juan county. Some of those beds are said to be forty
feet in thickness; these reports have not yet been verified.

The largest vein of coal in the world is at Bruhl, near
Cologne; it has attained a thickness of 340 feet in some places
—it is a brown coal.

Of the coal deposits of the world, New Mexico has been so
generously provided, that many other portions of the globe
lack much or all of their equity.

It is not necessary to go far from the limits of the territory
to find this disparagement; for Arizona and the Republic of
Mexico have been unduly discriminated against, in this
particular, in the economy of nature.

The estimated coal area of the world is approximated at
4,650,000 square miles. China is credited with 4,000,000 of
these, or more than six times the rest of the coal area of the
globe.

Next to China comes the United States with 280,000 square
miles; France, 2,086 square miles; Germany, 1,770 square
miles; and Belgium 510 square miles.

The coal area of New Mexico cannot be definitely approximated at the present time; yet it will be safe to estimate the
area at 2,500 square miles.

Area, however, cannot always be relied on as a correct crite-
rion of value. The anthracite fields of Pennsylvania embrace
an area of only 468 square miles, yet these, undoubtedly, are
of more value than any other like coal area in the world.

The world's production of coal for 1902, according to the
United States Geological Survey, was 884,795,343 short tons.

Of this amount the United States is credited with 301,582,-
348 tons; this is a little more than one-third of the entire coal
supply of the world.

It is difficult to conceive of the magnitude of the production of coal in the United States. To illustrate, suppose that twenty-two tons, the average car load, are placed in a car; this would give 13,708,288 cars. Allowing forty feet, which includes couplings for the length of each car, this would make up a train 103,850 miles long, sufficient to girdle the globe more than four times at the equator.

The amount of ore mined in New Mexico, in 1903, reached a total of 1,359,530 tons; this would form a train of 61,797 cars, 468 miles long; lacking only twenty-nine miles of reaching from Raton to El Paso over the Santa Fe tracks.

From the Annual Report of Jo E. Sheridan, U. S. Coal Mine Inspector, for 1903, the production of New Mexico by counties is as follows:

Name of County.	Output in tons for fiscal year ending June 30, 1902.	Output in tons for fiscal year ending June 30, 1903.	Increase in tons.	Decrease in tons.
Colfax.........	264,364	558,805	294,441	....
Lincoln.............	121,902	98,096	....	23,800
McKinley...........	561,436	539,910	....	21,526
Rio Arriba........	50,600	39,100	....	10,900
Santa Fe.........	103,461	92,359	....	11,102
Socorro	29,681	29,460	....	221
San Juan ...	1,500	1,800	300	....
Total	1,132,944	1,359,530	294,741	67,549
Deduct decrease..	....	....	67,549	
Net increase.....	....	....	227,192	

The percentage of increase of the year 1903 over that of 1902 is a very small fraction less than 10 per cent.

Colfax county stands first in production and has a coal area almost as great as San Juan or McKinley county. On account of the coking qualities of much of this coal it ranks among the best in the Territory. The output of this county in 1903 is seen to be more than double that of the preceding year.

The names of the mines operated during 1903, in the Territory by counties are as follows:

Colfax county—
    1.—Willow.
    2.—Dutchman.
    3.—Climax.
    4.—Sugarite.
    5.—Llewellyn.
    6.—Turner.
    7.—Dawson, No. 1.
    8.—Dawson. No. 2.
    9.—Dawson, No. 3.

Santa Fe county—
    1.—Cerrillos bituminous.
    2.—Cerrillos anthracite.
    3.—Anthracite "B" No. 33.
    4.—Block.
    5.—Pecos.
    6.—Uña de Gato.

Socorro county—
    1.—Hilton
    2.—Government
    3.—Bernal
    4.—Emerson

San Juan county—
    1.—Thomas
    2.—Morgan
    3.—Stevens

Lincoln county—
    1.—Old Abe
    2.—Capitan, No. 1
    3.—Capitan, No. 2

McKinley county—
    1.—Gallup
    2.—Weaver
    3.—Catalpa
    4.—Clark
    5.—Otero
    6.—Thatcher
    7.—Rocky Cliff

        8.—Union
        9.—Black Diamond
        10.—Casna
        11.—Gibson (resumed in June, 1903).
Rio Arriba county—
        1.—Monero, No. 1
        2.—Monero, No. 2
        3.—McBroom
        4.—Kutz

Beside the regularly operated mines there are numerous prospect pits in several of the counties, which supply coal in a small way, and are not classed in the producing list.

The coal measures of New Mexico were all laid down during the Cretaceous period; this period, perhaps, will include more than one-third of the rock surface of the territory. The formations laid down during the whole of Cretaceous times, in New Mexico, will approximate a total thickness of 5,000 feet. The most important series of this great rock system, from an economical standpoint which concerns the territory, are the later or uppermost portions, known as the Fox Hills group of the Upper Montana and the Laramie series.

To these two series belong the entire coal measures of the territory. It now seems certain that the Madrid deposits* south of Cerrillos and the Carthage field belong to the Fox Hills series, which horizon is immediately below the Laramie. No coal has yet been found in New Mexico belonging to the Carboniferous period. From observations by the writer there are a number of deposits in the territory which may be proven to exist in the Fox Hills horizon, that have been heretofore considered in the Laramie.

The extraordinary condition found at the Madrid field is scarcely paralleled in any other region on the globe. Here are four distinct workable veins of anthracite which are the nearest to the surface; below these are several workable veins of bituminous coal. It seems that these conditions were effected by intrusive dikes or laccoliths in proximity to the coal. Since anthracite is nothing more than metamorphosed lignite or bituminous coal, it is always expected to

---

*The Geology of the Cerrillos Hills by D. W. Johnson, page 40.

Fig. 37.—OPENING AT THE PINAVITITOS COAL MINE.

find associated intrusives in the immediate vicinity of such deposits or coking coal.

A section of the Madrid fields shows, beside the four anthracite veins, twelve others which may be eventually worked. At Gallup there are from four to five veins ranging in thickness from three feet to six and one-half feet. Similar conditions exist in most all the other beds in the territory.

The new field now being opened up at Hagan (*Uña de Gato*) in southeastern Sandoval county, by the New Mexico Fuel and Iron Company, has every promise of being one of the best producing districts in New Mexico.

New coal fields and mines are being found and opened up every year. Preparations are being made to open up the Sloan field east of San Felipe, in Sandoval county; this is virtually the same field as that of the New Mexico Fuel and Iron Company. Near Gallina postoffice in Rio Arriba county, a new field has been recently discovered in 1904 by R. W. Tandy and W. H. Raymer.

The vast coal areas in western Socorro and Valencia counties and most of those in San Juan county have never been touched. Near the Ladrone mountains in Socorro county some new fields have lately been located and surveyed.

The following table shows the statistics of the coal mining industry for the fiscal year ending June 30, 1903, taken by counties:

Counties.	Amount used in operating Mine.	Net product.	Estimated value of net product for the fiscal year ending June 30, 1903.	County percentage of total output of New Mexico.
Colfax.	10,753	548.052	645,213.70	40.367
Lincoln.	1,000	97,096	195,240.00	7.217
McKinley	10,580	529,330	671.536.37	39.712
Rio Arriba.	1.000	38,100	58,000.00	2.956
Santa Fe.	9,019	83,340	184,474.73	6.793
San Juan		1,800	2.445.00	.132
Socorro	375	29,085	38,319.00	2.159
Totals	32,727	1,326,803	1,795,208.80	99.336

Production of coke in New Mexico for the fiscal year ending June 30, 1903. (From report of U. S. Coal Mine Inspector, Jo E. Sheridan).

Name of oven	Number of tons	Estimated value per ton of 2,000 pounds	Value of product at the ovens
*At ovens of the Dawson Fuel Co., Dawson, Colfax county ...	18,074	$3.00	54,222.00
†At ovens of the Raton Coal and Coke Company, Gardiner, Colfax county.......	8,279	2.50	20,697.50
Total	26,353	....	74,919.50

*The coal from which this coke was made was mined from the Dawson mines.
†The coal from which this coke was made was mined from the Wilson mines, owned by the Raton Coal and Coke Company.

Analyses of New Mexico Coals.

(Taken principally from Jo E. Sheridan's Report, U. S. Coal Mine Inspector)

McKinley County.

Analysis of coal from Catalpa mine, near Gallup:

(Owned and operated by American Fuel Company.)

Moisture..........	6.66%
Volatile Matter.	40.13
Fixed carbon	45.56
Ash.......	7.65
Total....................	100.00%

Analysis of coal from Weaver mine, at Gibson, near Gallup:

(Owned and operated by Colorado Fuel and Iron Company).

No. 3 seam—

Moisture........................	9.13%
Volatile matter......	38.45
Fixed carbon	49.43
Ash..............................	2.99
Total	100.00%

No. 5 seam—

Moisture..	8.23%
Volatile matter.	40.61

Fixed carbon........................ 45.17
Ash.... .... 5.99
                                   _____

        Total 100.00%

The Gallup mine is being operated upon the same coal
seams as the Weaver mine, viz, No. 3 and No. 5, and analyses
are similar to those given above for those seams.

Colfax County.

Analysis of coal and coke produced from Raton Coal and
Coke Company's mines at Raton:

Coal from Raton Coal and Coke Company:
        Water....... .... 0.75%
        Volatile matter.... ... . 34.40
        Fixed carbon 56.93
        Mineral ash. 7.92
                                          _____

        Total 100.00%
                                          _____

        Coke 64.85
        Character of coke, very strong and
        tough; color of ash, very light ocher;
        character of ash, soft and light.
        Sulphur (as sulphide) .... .016
        Sulphur (as sulphate)........... ... .022
        Phosphorus........................ .014
        Specific gravity......... .... 1.291
        One cubic foot weighs in pounds..... 88.690

Analysis of Mineral Ash:

        Silica....... ... . 44.16%
        Alumina................. .... 39.28
        Oxide of iron........ .... 2.95
        Calcium oxide. 7.41
        Magnesium oxide................... 3.27
        Sulphate of calcium .... .41
        Alkalies and loss 2.52
                                          _____

        Total 100.00%

Analysis of coal produced from Raton Coal and Coke Company's "Willow" mines at Van Houten:

        Moisture........................... 3.61%
        Volatile matter. 35.55
        Fixed carbon....... .... 51.73
        Sulphur .... .63
        Ash........ .... 8 48
                                          _____

        Total 100.00%

Analysis of coal produced from the Raton Coal and Coke Company's "Dutchman" mine at Blossburg:

Moisture	1.28%
Volatile matter	33.90
Fixed carbon	56.68
Sulphur	.65
Ash	7.49
Total	100.00%

Analysis of coal from the Dawson Fuel Company's mines at Dawson:

Moisture	1.32%
Volatile matter	37.47
Fixed carbon	52.50
Sulphur	.21
Ash	8.50
Total	100.00%

Santa Fe County.

Analysis of coal from the Cerrillos bituminous mine of the Colorado Fuel and Iron Company at Madrid:

(W. D. Church, analyst, December 2, 1893.)

Water	2.00%
Volatile matter	39.00
Fixed carbon	53.76
Mineral ash	5.24
Total	100.00%
Coke	59.00

Character of coke, strong and tough; color of ash, light yellowish gray; character of ash, soft and light.

Sulphur (as sulphide)	.010
Sulphur (as sulphate	.022
Phosphorus	.006
Specific gravity	1.410
1 cu. ft. weighs in lbs	88.135

Analysis of mineral ash:

Silica	26.93%
Alumina	32.41
Oxide of iron	3.96
Calcium oxide	24.68

Magnesium oxide...................... 10.32
Calcium sulphate .... .21
Alkalies and loss........ .... 1.49

Total 100.00%

Analysis of Cerrillos anthracite.

(Analysis furnished by the Colorado Fuel and Iron Company.)

Volatile combustible matter 3.18%
Fixed carbon 88.91
Water 2.70
Ash............................... 5.21

Sandoval County.

Analysis of coal from the Uña de Gato mines, operated by the New Mexico Fuel and Iron Company-

(Jas. O'Hardy. analyst, 1902.)

Moisture.......... .... 6.25%
Volatile matter 40.40
Fixed carbon 47.56
Ash............................. ... 5.79

Total................. .. .. 100.00%
Sulphur.............................. .62

Lincoln County.

Analysis of coal from the New Mexico Fuel Company's mines at Capitan:

Water........... .... .75%
Volatile matter...... .... 41.25
Fixed Carbon........ .... 47.00
Ash......... .. . . 11.00

Total........................ 100.00%
Sulphur...... .... .735

R. C. Hills, the geologist of the Colorado Fuel and Iron Company, who examined the property of the New Mexico Fuel Company, constructed a coke oven of adobe bricks and coked some of the coal from the Akers seam, which gave the following analysis:

Water 1.450%
Volatile matter................... ... 3.900
Fixed Carbon...... .... 76 825
Ash.... .... 17.825

Total.... .... 100.000%
Sulphur............... .... .611

Socorro County.

Analysis of coal from Hilton's mine at Carthage:

(Fayette A. Jones, analyst, March, 1902.)

Moisture............ ....   .06%
Volatile matter.... .... 37.55
Fixed Carbon........ .... 54.88
Ash................................. 7.51
                                    _____
        Total 100.00%
Sulphur........... 83

Analysis of coal from the Emerson mine at Carthage:

Moisture........................... 1.00%
Volatile matter................. .... 39.40
Fixed carbon.............. .... 53.20
Ash 6.40
                                    _____
        Total................ .. .. 100.00%

CHAPTER XXX.

SALT.

But his wife looked back from behind him, and she became a pillar of salt. Genesis 19:26.

The saline lakes of New Mexico must be regarded as an important factor in the resources of mineral wealth of the territory.

These saline deposits have supplied the culinary wants of the aborigines throughout this part of the southwest from the earliest times.

The early Spanish explorers when enumerating the products of the country have on many occasions mentioned salt.

There can be little doubt about these saline deposits having played an important part as a food savor of the ancient Pueblos, centuries before the discovery of America.

Salt may properly be regarded as a civilizer; we find it was extensively used by the Pueblo Indians of the southwest, the ancient Mexicans and Peruvians. It should also be noted that the savages of the western plains, and cannibals rarely ever or never tasted the article. The docile effect which salt has on wild animals is well known to all; and there is not a civilized people today who do not use the article in quantity.

Experiments have shown that a solution of chloride of sodium stimulates the action of the heart and apparently sustains life; this may account for the fact that the primordial types of life, necessarily appeared first in the sea. The use of this simple substance may also be responsible for the increasing longevity of the human race; and its general absence from the food of most animals would explain their shortness of life.

Statistics show a steady increase of the production of salt; the United States now stands at the head of the list of salt producing nations. The output of the United States in 1902 was 23,849,221 barrels of 280 pounds each, valued at $6,617,-

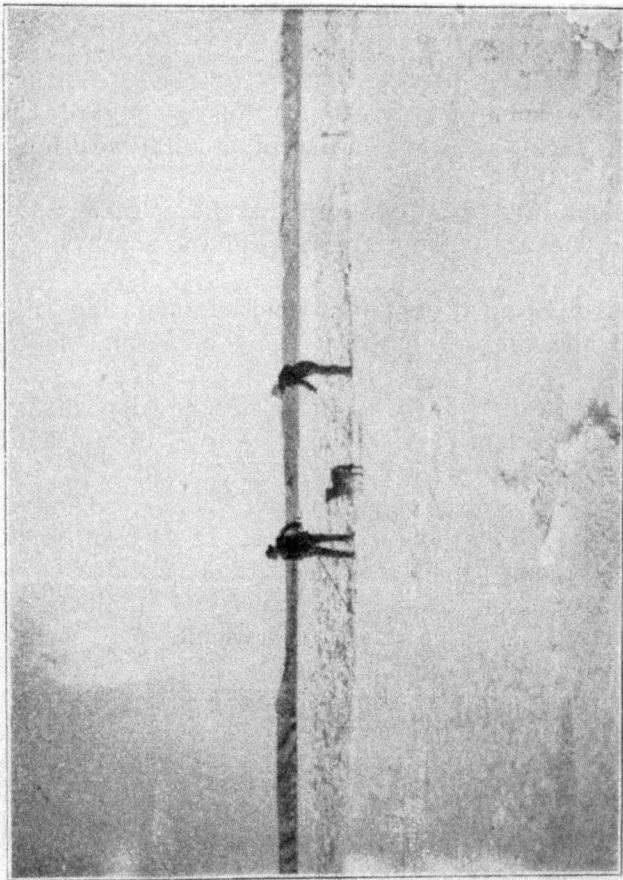

Fig. 38 - SHOWING METHOD OF COLLECTING SALT AT BIG SALT LAKE, ESTANCIA PLAIN.
Photographed by F. A. Jones, December, 1903.

449.   Michigan now ranks first among the states in production, having an output of 8,131,781 barrels in 1902.

Up to the present time this territory has never manufactured or refined salt on a commercial scale.   The saline lakes, however, furnish a large amount of the crude product which is used by the stockmen for salting their herds.

This salt contains many impurities, being collected by scraping up the deposit which forms in a thin layer over the surface of the lake bottom, during the dry season.

The geology of the saline deposits, in New Mexico, is the same and would apply in a general way to like deposits that are in progress of formation in other portions of the globe.

The association of salt with gypsiferous beds is discussed under the caption of Gypsum Cement Plaster.   As there pointed out, both gypsum and salt are closely related to the "red beds" or Permian formation.   It may be taken as a guide, in looking for deposits of salt and gypsum in New Mexico, to first select the horizon in the red beds series; then at or near their base or some associated basin prospect for those minerals.

Having determined the salt and gypsiferous horizon it is an easy matter to account for the saline marshes and lakes with their attendant deposits.   Saline and alkali solutions collecting in impervious basins and after extreme desiccation of the water, will deposit their solid materials when the point of saturation is reached; the gypsum is always precipitated first.   Big Salt Lake of Estancia plain with its adjacent basins has now reached the period of excessive old age.

What now remains of those saline basins and gypsiferous dunes, is nothing more than the lingering of another White Sands plain and is only a little farther advanced in age.

The forces of the winds which dug those saline and alkali graves to their greatest depth (perhaps 60 feet below the uppermost strata of the ancient lake bed of Estancia plain) are now at work rapidly refilling those burial pits with the same material that it excavated ages ago.   Should this deposit not be wrested from the encroaching sands of the plains, in time it would add another reserve to the world's inexhaustible supply of hidden treasure.   Thus the processes of nature are continuous and ever at work; renovating the old into the new and aging the new from the mantle of the past.

The saline and alkali lakes of Estancia plain necessarily occupy the lowest point in the plain which lies between the Trinchera Mesa and the Manzano Mountains.

Big Salt Lake is the only one of this group that produces salt to any extent. This particular lake is destined to play a very important part, in the near future, in supplying salt to the people of the territory. This property has recently passed into the hands of the Pennsylvania Development Company and is within a few miles of its recently constructed railway, the Santa Fe Central. It is understood that a plant will be erected within the very near future for the purpose of refining and purifying the crude salt.

The waters from these lakes are saturated with salt and other soluble matter forming a very strong brine. Dr. C. R. Keyes had a gallon of water from the Big Salt Lake analyzed at the New Mexico School of Mines, with the following result:

Constituents.	Grains per Gal.
Sodium Chloride (Common Salt).......	10,900
Potassium Sulphate.............	7,103
Magnesium Chloride.............	4.192
Magnesium Sulphate	3,004
Alumina.............	64
Silica................	35
Volatile matter	4,102

It is seen by this high percentage of solid matter that the water is almost completely saturated and on standing in a vessel for a few hours will deposit crystals of salt on the sides and bottom of the vessel.

Lying in the soft black mud which covers the bottom of the lake are large crystallized chunks of the mineral bloedite —a hydrous double sulphate of soda and magnesia—some pieces of which will weigh several hundred pounds. This is at present the only place in the United States where this rare mineral is known to exist.

The Salt Lakes of the "White Sands" plain furnish about the same geological conditions as those noticed at the Estancia plain lakes. Nature has in this case operated on a grander scale and her work is not quite so far advanced.

From a visit to this curious plain it was found that a saline horizon consisting of the red beds series, almost encircled the basin. Lying near or at the foot of these red beds, was a salt area consisting of salt flats, lakes, springs and marshes; these saline waters and deposits lay concentric to the escribed series of the red beds. The central portion of the plain is composed principally of gypsum dunes. Near the south end of the great flow of basalt (locally known as the *mal pais* and which lies in the lowest part of the plain to the north), is a small salt lake which supplies the ranchmen with salt in that section of the country. Another salt lake may be found near the southeast side of the White Sands, which is of much importance to the ranchers for miles around.

The several salt lakes in the "White Sands" basin have been patronized for centuries by the aborigines.

Dr. C. L. Herrick in a bulletin of the University Geological Survey of New Mexico, in 1900, reports the following analysis of salt from this region:*

Sodium Chloride....................	90.83%
Sodium Sulphate................	1.57
Calcium Sulphate.............	.65
Potassium Carbonate...........	1.25
Magnesium Chloride....	Trace
Sandy Impurities......	5.70
Total........................	100.00%

Dr. Herrick also reports on a sample of brine from the "White Sands" region:

	Grains per gallon.
Sodium Chloride...................	8,450
Magnesium Sulphate	316
Calcium Sulphate.	465
Sodium Sulphate................	452

The salt flats all contain sodium and potassium, sulphates and carbonates and some magnesium chloride, as well as a large percentage of sodium chloride.

The local commercial value of these lakes to the stockmen, as a source of supply of salt for their herds, is inestimable.

*Dr. C. L. Herrick describes the "Geology of the White Sands" in the Bulletin of the Hadley Laboratory of the University of New Mexico, Vol. II, Part I. Published in the year 1900.

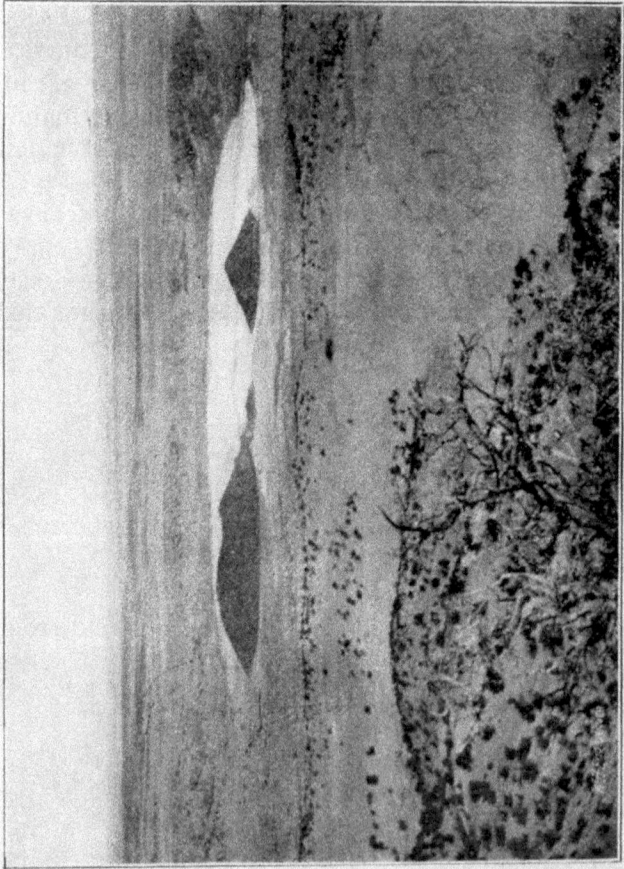

Fig. 39— CRATER SALT LAKE, WESTERN SOCORRO COUNTY. Photographed by F. A. Jones, April, 1903.

It is possible that these deposits may be worked on a commercial scale before many years elapse.

The Zuñi Crater salt lake in western Socorro county is by far the most interesting of any of the sources of supply of salt in New Mexico.

The distance across the main depression or circular basin, from wall to wall, at the crest, is one and one-fourth miles. Standing near the center of the basin are two basaltic cinder cones, perhaps 150 feet in height each; the larger cone to the southwest has its conduit still open; the smaller cone to the northeast is solid. Evidently from the nature of the two cinder cones, the upward bulging of the strata and from the comparatively little massive basalt in the neighborhood, the volcano must have been of the explosive type, which burst through the Cretaceous covering apparently instantaneously. The ejecta to the north and east, consisting mainly of a comminuted breccia and mud, would indicate vulcanism in presence of much water. The cinder cones themselves, with the attending travertine at the base of the conduit cone facing the solid cone, would go to substantiate the theory that it was a "fight to the finish" between fire and water.

The theory of the saline character of this crater lake as advanced by Dr. Herrick, that the open conduit extends downward to the underlying red beds series, is no doubt correct; since the red beds outcrop just to the north and south of the crater, and no doubt extend under the Cretaceous rocks which embrace the crater basin.

The Zuñi Indians have always associated this crater lake with many superstitions; they have a tradition that this volcanic crater appeared in a single night. Is it probable that the earliest Zuñi Pueblos witnessed the last convulsions of this now extinct member of Vulcan? The ascending waters in the crater conduit carrying in solution the saline matter of the red series, diffuses into the salt lagoon where its burden is precipitated at the bottom of the lake. The salt is shoveled from the lake into small flat boats by Mexicans, after which it is pilled up on the bank ready for marketing. This lake supplies all the surrounding ranches with salt within a radius of one hundred miles or more. The quality of this salt is the best in the territory. Since the process of salt-making is continuous at Crater Salt Lake, no fear of exhausting the supply need be entertained.

Should the feed supply of the lake be cut off, the waters alone in the lagoon contain approximately 500,000 tons of salt. Beside the salt contained in the brine, several strata of pure salt varying from three inches to a foot thick are reported to exist in the bottom of the lagoon. It is quite probable that some of the lower strata of salt extend entirely around the cinder cones and spread over the entire floor of the older lake bed; these deposits having been covered by silt by the gradual filling up of the lake basin. In all probability there are several millions of tons of salt which could be made available. No analysis of the lake and spring waters of this salt basin has ever been made, although the writer selected a sample from the Crater Lake for analysis, which he intended to make while Director of the New Mexico School of Mines, but press of school duties prevented.

The above saline lakes are the most important which occur in the territory; there are numerous saline flats which may eventually yield salt after their proper exploitation. It is learned that in drilling a well in the southwestern part of the San Augustine plain, near the fresh water holes, below Fullerton's ranch some five miles, that considerable salt water was encountered, which stock refused to drink, and the well was abondoned.

East of the Pecos, in Eddy county, are valuable saline lakes.

CHAPTER XXXI.

CEMENT, PLASTER AND LIME.

Of the many mineral products of New Mexico that have economical value, and exist in unlimited quantities, are the materials from which cements, plasters and lime are made.

The finished products manufactured from the raw materials will ever be on the increase, in order to keep pace with the growing demands and needs of a progressive civilization.

Owing to the world's supply of wood, copper and iron becoming practically exhausted within less than half a century, should the rate of demand proportionately increase as it has done during the last half century, these three materials will necessarily be largely eliminated from present uses within a few years on account of increased cost. Therefore, in order that equilibrium be restored, the uses of cement materials will evidently increase as the prices of wood and the commercial metals advance. It is no idle dream, and the future is not far distant, when the commodities of gypsum from the west will be marketed in the east. The increasing popularity of mission architecture in the west, will naturally greatly increase the local manufacture of a good cement plaster, which can be had at a moderate cost.

Cement-concrete ties have been tried on the Pere Marquette Railway, and also on the Wabash railroad, with satisfactory results.

Cement or cement block houses have the advantage of being more durable than wood, as well as being fireproof and excel in sanitation.

Portland Cement.

Perhaps, nothing could be utilized to greater advantage or extent of New Mexico's natural resources, than the material from which cement is made.

By the word cement, is meant that kind, in particular, having hydraulicity* properties—that is, having the power to se or harden under water.

In this age the uses of cement register on the scale of progress the degree-of advancement attained by the people in any particular locality.

Its uses are so diversified in all the important and economic works of engineering and architecture that it is destined, in a great degree, to supersede wood and iron. It possesses important qualities superior to both those of wood and iron, since it will neither decay nor oxidize, but becomes harder and more durable with age, when properly prepared.

It is a remarkable achievement of man that he can manufacture within a few hours a rock more durable than is possible for nature to duplicate in millions of years.

The use of cement of some types antedates history. There are samples of fairly good cement mortar in the pyramids, in the pavements of Pompeii and Herculaneum. This cement used by the ancients was a natural product, chiefly volcanic ash or earth, and is very much inferior to the modern article.

"Trass," a volcanic earth is found in some parts of Holland on the Rhine, and in certain localities of New Mexico. Pozzuolana, a light or dark colored, porous lava or ash is the material from which the Romans made their cement. Pozzuolana was so named from its proximity to the town of Pozzuoli, Italy, which is situated near the base of Vesuvius. This Pozzuolana was the first substance that attracted the attention of modern observers on account of it possessing the peculiar property of hydraulicity. Adobe mud, so extensively used throughout the southwest, possesses this peculiar property in a small degree, only.

The Latin race used a hydraulic cement, which was made from Pozzuolana with perhaps a certain admixture of clay and lime; and it appears that the art of manufacture was lost during the dark ages.

This cement was of a very quick setting kind, and would solidify in a mass when thrown into a running stream.

Robert Smeaton, in 1756, is givin the credit of reviving this

---

*Hydraulicity is a word newly coined and may be defined as a property which certain materials possess of setting or hardening under water.

lost art; the product made by him was supposed to be identical to that made by the ancient Romans.

Parker, Wyatt & Co. of England, were the first to take out a patent on artificial cement; this was in the year 1796.

Vicat, a French experimenter, first observed and noted the principle of hydraulicity in the year 1818.

Portland cement was made from the natural limestones by Joseph Aspdin of Leeds, England, in 1824. It, however, remained for Pasley, in 1825, to obtain a cement by the burning of river mud from the Medway, which was impregnated with the salts of sea water and lime. This gentleman is properly the true founder and discoverer of the modern Portland cement.

Good Portland cement has about the following analysis:

Lime....	61.00%
Silica...............................	23.50
Alumina............................	8.00
Iron oxide.........................	3.00
Magnesia	1.00
Soda and potassa oxides	1.25
Sulphate of lime....................	2.25
Total	100.00%

Portland cement receives its name from its similarity of color to the Portland oölitic lime stone of Dorsetshire, England.

The stone found in nature from which the original Portland cement was made, contains about 21 per cent of clay and 79 per cent of lime-carbonate. The clay is composed of about one part of alumina to two parts silica. When roasted at a high temperature for a considerable length of time, most all the alumina and silica combines with a part of the lime; the uncombined lime of the product is not sufficient to cause the same to slack in the presence of water. The roasted product is reduced to a powder by grinding.

The superiority of Portland cement over that of other kinds is supposed to be due to the double silicate of aluminum and lime, which is formed only at a high temperature.

The density of cement is greatly increased by long continued roasting, which materially adds to the strength of the product; provided, the point of vitrification is not reached.

The Rosendale type of cement is very extensively used in the United States and is manufactured in many places, viz: Utica, Ill., Kansas City, Fort Scott, Kan., Milwaukee, Trinidad, Colo., Louisville, Ky., and various other parts of the country: it is made from an argillo-magnesium limestone. The stone is quarried and then crushed to a suitable size, afterwards it is calcined at a certain temperature and then ground to a very fine powder. This cement is rather quick-setting and does not have the strength of the slower-setting kind, of which the Portland is the best type.

Rosendale cement is frequently termed Roman cement or natural cement since it is made by calcining the natural stones which contain all the necessary constituents for that class of cement.

Rosendale cement was first manufactured in this country in 1837; but Portland cement was not made in the United States until 1875.

Pennsylvania has the honor of constructing the first Portland cement plant in the United States; the product the first year was 1,700 barrels.

Perhaps, twenty-four twenty-fifths of the Portland cement manufactured in England and the United States is now made by artificial admixture of the several necessary constituents.

The mass is roasted at the proper temperature and then ground to a very fine powder. This powder would all pass through a 625-mesh sieve and about 20 per cent should be left on a 4,900-mesh sieve; this degree of fineness is supposed to give the best results.

The large and numerous beds of marls found in various parts of the Territory, which once formed the bottoms of ancient lakes, will furnish the necessary raw material for the best grade of Portland cement.

The marl beds of certain portions of Estancia plain, along the line of the Santa Fe Central Railway, would by reason of transportation facilities and superior quality of the marls, make an ideal place for the establishment of a hydraulic cement industry.

The beds of cretaceous shales in Colfax county at Springer, make a splendid cement; a plant was erected at that point several years ago; but, from some unknown reason, was closed down indefinitely, shortly after its completion.

This important product, so necessary in building and so extensively used in all engineering and architectural works, is used but comparatively little in New Mexico.

The finished product is imported into the Territory at a very great cost which largely curtails its uses.

Why should New Mexico people be like the early Dutch of New York, who would send to Holland to buy brick, when the material could be had at home?

The graphical illustration gives a clear conception of the tremendous growth of the Portland cement trade in America since 1890. The figures at the top of the diagram represent years and those to the left are based on the barrel as a unit.

Gypsum Cement Plaster.

Gypsum must be considered as a coming factor in the economic distribution of the mineral wealth of New Mexico.

Its uses are manifold and varied in character: such as cement plaster (which has fire-proof qualities), stucco, dental plaster, filler in paper, fertilizer, scagiola finish (an imitation of onyx and marble), cement paint, plaster of Paris, building blocks, etc.

The virgin deposits of this useful material in New Mexico have been, as yet, practically untouched. These beds frequently attain great thicknesses and cover large areas.

With our meager knowledge of the full geological distribution of this material throughout the territory, it would be absurd at the present time to attempt to estimate its extent or quantity.

Deposits of salt are always associated with gypsum; but, a deposit of gypsum does not necessarily imply that it is associated with a bed of salt. The reason of this is obvious, since gypsum is the first substance to crystallize out of a supersaturated brine; before further desiccation, which would be necessary for the deposition of salt, the water from some cause would become fresh, or the salt already deposited, would again pass into solution. In most all the deposits of gypsum in the territory where the beds have been un-

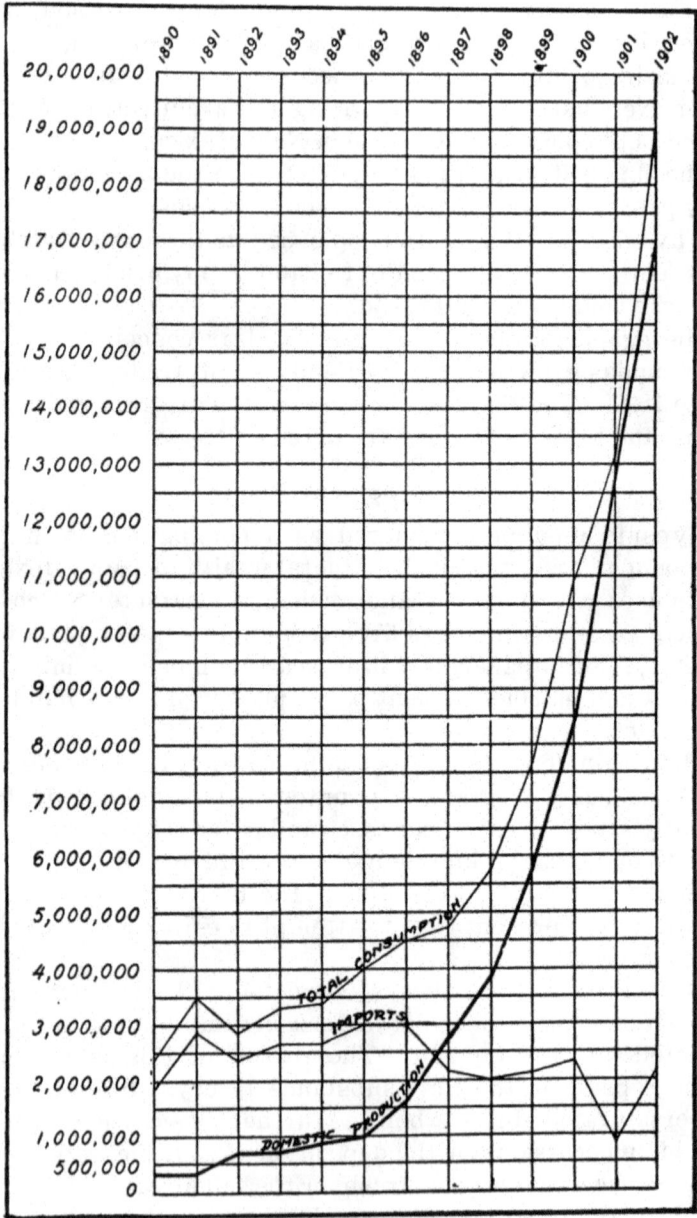

Fig. 40 -GRAPHICAL DIAGRAM OF THE CEMENT INDUSTRY
OF THE UNITED STATES.

disturbed, the stratifications, due to water, are plainly to be be seen.

It is, therefore, evident that in most instances the gypsum deposits of New Mexico were laid down in salt water lakes or arms of the sea that became separated from the main body of ocean water. It further appears, that all of the extensive beds of gypsum of the territory are an accompaniment of the Permian or red beds formation. Since the Permian epoch was one of general unrest, there were oscillations up and down of the existing land areas which disconnected and reconnected successively, arms of the sea, making favorable the conditions necessary for the deposition of gypsum.

The gypsiferous beds in most places show that such movements have taken place; for strata of limestone and gypsum would correspond to a period of elevation and desiccation of the inland lakes; while the deposition of shales and sands would correspond to the time of downward movement and the disturbing force of rushing waters from the ocean. This explanation would fully account for the various stratas of impurities which occur in the gypsiferous deposits throughout the southwest and elsewhere.

There are some deposits of gypsum, in New Mexico, where the beds do not have a stratified appearance, which would indicate a different mode of deposit. An example of such deposits may be found at the Jones' iron mines in the eastern part of Socorro county at the north end of the Sierra Oscura, near the mine cabin. Some of those beds are very white and are good examples of alabaster, having a fine granular texture. Their occurrence, in part, at least, would appear to be due to the oxidation of pyrites which is associated with the monzonite dike that ruptured the Carboniferous limestones.

The oxidation of the sulphur would produce sulphuric acid and this in turn would attack the contiguous limestones, thereby forming gypsum. Owing to the enormous beds of gypsum at this place, this theory should be excluded from consideration, unless we would suppose the entire bodies of massive iron ore, were at one time the sulphides of iron. The bedded structure of the iron deposits would scarcely admit such a conclusion.

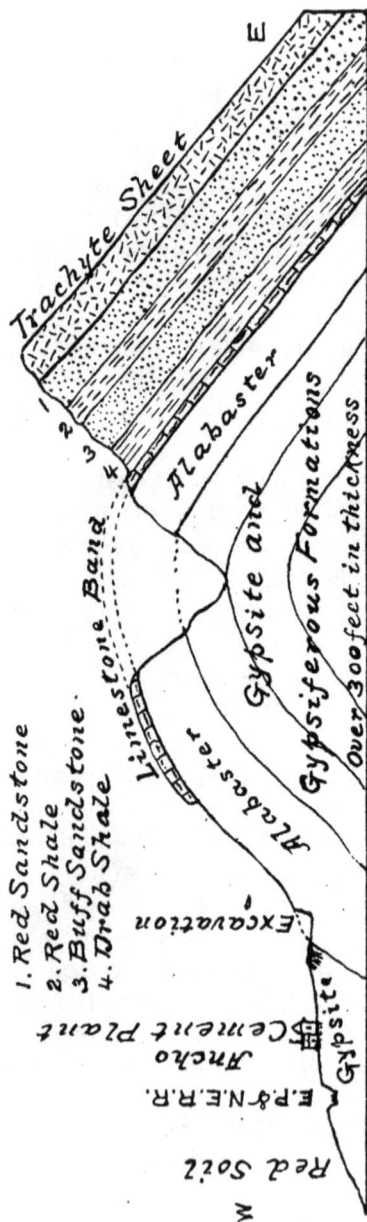

Fig. 41—GEOLOGICAL SECTION OF ANCHO GYPSUM DEPOSITS.

The protruding of this enormous molten or plastic mass was evidently attended by much chemical activity.

Sulphuric acid vapor would naturally be an accompaniment of these igneous disturbances and would readily attack the overlying limestones which were pushed upward from their beds and much shattered by the protruding dike.

The checking of the limestone more fittingly prepared it for the assault and action of the sulphuric acid vapors; which resulted in forming those agglomerated beds of gypsum.

The reaction of sulphuric acid acting on limestone in the presence of water can be best shown by the chemical equation:

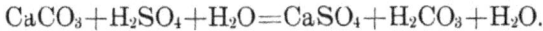

$$CaCO_3 + H_2SO_4 + H_2O = CaSO_4 + H_2CO_3 + H_2O.$$

Since the carbonic acid is not a stable compound it breaks down into carbon dioxide and water.

Therefore,

$$CaSO_4 + H_2CO_3 + H_2O = CaSO_4 + 2H_2O + CO_2.$$

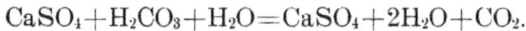

This gives in the last member of the equation, Gypsum ($CaSO_4 + 2H_2O$), with its water of crystallization and liberates the carbon dioxide (CO_2) gas.

The deposits lie on either side of the monzonite dike and, perhaps, more than three-fourths of the original beds have been removed by the actions of wind and water.

Doubtless, there are a number of other deposits of gypsum in the Territory which have been formed under similar igneous and volcanic conditions, as well as by the action of acidulated waters of thermal springs on calcareous tuffs and tufas. However, the point should not be overlooked, that the principal number and most extensive deposits of gypsum which occur in New Mexico owe their origin to the desiccation of salt water lakes and bays which were formed inland or were cut off from the sea.

A geological cross-section of the extensive gypsum beds of the Rock Island Cement and Plaster Company at Ancho, Lincoln county, is here given.

These beds are of enormous areal extent and are of great thickness. A well was bored for water by the El Paso and Northeastern Railway Company on the west side of its tracks at Ancho. The well was sunk to a depth of over four hundred feet and was abandoned as no water was found.   It is said that the boring was in a gypsiferous formation the greater portion of the depth.

These deposits are of special importance, since a plant of one hundred tons capacity has recently been installed and fine products of stucco, cement, cement plaster, plaster of Paris and dental plaster are made.

The gypsum occurs in two forms, locally known as gypsite and alabaster.   The gypsite is a gray or yellowish deposit containing a moderately low per cent of silica with some organic impurities.   The massive alabaster variety lies in stratified beds dipping about thirty degrees toward the west and has a whitish gray appearance with some small dark streaks or zones running through it; these dark streaks are principally organic impurities, which disappear on calcination.

The gypsite has the appearance of a re-deposit which has come from the erosion of the original gypsiferous beds above.

The analysis of this gypsite, which was made by M. Carleton Ellis, S. B., of Boston, who is a specialist on plasters and cement, is here given:

Sulphate of lime	63.95%
Carbonate of lime	20.04
Silica	3.57
Oxides of iron and aluminum	2.01
Magnesia	.89
Chlorine	.09
Moisture	9.45
Total	100.00%

Brown cement plaster is made from this kind of material.

The Rock Island Cement and Plaster Company, manufactures cement plasters of various shades of color; such as chocolate, brown, light yellow, reddish brown, gray and white.

The gypsite from which the cement plaster is made, comes from a gypsiferous bed which is apparently a re-deposit; not being nearly so pure as the solid beds of alabaster.

The finer grades of plaster, such as plaster of Paris, stucco and dental plaster, are made from the alabaster deposit.

This stucco is scarcely excelled by any like product in the

world; it is a close rival in quality to the celebrated Nova Scotia article.

Some of the dental plasters and stuccos will set in less time than three minutes; on account of this rapidity in setting a retarder is usually introduced.

The technology of preparing the various plaster cement products is simple, though much experience and judgment is required. The solid raw material is crushed in a jaw-crushing machine, similar to an ordinary ore crusher.

From the crusher the fragments fall into a rotary grinder, something on the order of a coffee mill, where the product is still further reduced in size. A bucket elevator lifts these small particles to a point where the same is fed from a spout into an ordinary buhr stone mill, where the material is reduced to a very fine powder. This floured gypsum, after its exit from the buhr mill, is passed into a second elevator and hoisted to a bin over the calcining kettles. From this bin the fine gypsum is run slowly into the kettles. The temperature of the kettles is kept at 212° F. or 100° C. The gypsum is constantly stirred by means of revolving paddles or arms, which keeps the material from caking and burning at the bottom.

In an hour or so the water has been driven off, the temperature is gradually raised to about 285° F. or 140° C. The product is hoisted to the upper story of the mill by a bucket elevator and is there bolted in a revolving screen similar to that used in a flouring mill. After the bolting operation, the process is complete.

The operation of treating the gypsite or cement plaster is somewhat curtailed from that of the process of preparing the finer grades of the white product; inasmuch, as the coarse crushing and bolting processes are not required.

The most delicate part of the process in cement plaster manufacture is in regulating or obtaining the correct temperature during the operation of calcining; such skill can only be acquired by practice. A trained calciner, can tell from the appearance of the material the proper degree of heat, independent of the reading of the thermometer. Different products in cement plaster manufacture, require different degrees of heat during the calcination process.

Fig. 42.—ANCHO CEMENT PLANT.

If gypsum is over-burnt, heated over 204° C., its property of hardening with water is destroyed; the cause is, perhaps, from it having been converted into anhydrite, which does not re-form with water.

The plant of the Rock Island Cement and Plaster Company was begun in September, 1902, and completed in the following spring.

Only one calcining kettle was placed in position at that time and operated during the summer of 1902. The business has justified the installation of another kettle, the same capacity as the first, increasing the output to one hundred tons per twenty four hours. This plant is similar in construction to many of those in Kansas.

Before leaving this subject it would seem proper here, in addition to the brief mention made of the plain of the "white sands" in the chapter on salt, to add a few more words about this marvelous gypsum deposit.

Indeed, the plain of the white sands may well be spoken of as one of the greatest wonders in the southwest. When viewed from the adjacent mountain tops, the unsuspecting person easily mistakes the expanse of the snowy field for water. This gypsiferous sea is about thirty five miles in extent north and south, and the greatest breadth at its southern extremity is approximately eighteen miles; the area is somewhat triangular in shape and embraces nearly 350 square miles.

Restless as the ocean waves are the ever shifting sands of this snowy waste; dunes are formed here to day and depressions take their place on the morrow.

An analysis of the sands made by the New Mexico College of Agriculture and Mechanic Arts, gives the following results:

Calcium sulphate (gypsum).	97.00%
Calcium carbonate............	2.06
Magnesium sulphate....	.12
Magnesium carbonate.	.06
Potassium sulphate................	.07
Sodium carbonate...	trace
Sodium chloride.....	trace
Total......................	99.31%

The water of crystallization is included in the calcium sulphate.

A concern in Alamogordo is hauling these sands to that town, where it is manufactured into a cement plaster, similar to that made at Ancho.

Slag Cement.

Slag cement is made as a by-product from the basic slags of iron furnaces and is a comparatively recent industry. There are now several slag cement plants in operation in Pennsylvania; the industry is growing quite rapidly.

The composition of slag suitable for the manufacture of this kind of cement, is about the following average composition:

Silica.........	32.72%
Alumina...........................	12.95
Iron...............................	2.51
Lime	47.67
Magnesia........	2.71
Sulphur	1.44%
Total.........	100.00

In the preparation of slag cement the material is coarsely crushed after drying, after which a certain amount of slacked lime is added; the whole is then thoroughly mixed and then reduced in tube mills.

The finished product has the following average composition:

Silica	27.33%
Alumina............................	11.61
Iron...............................	2.43
Lime...............	55.83
Magnesia...........................	1.93
Sulphur	.87
Total	100.00%

Slag cement is called Pozzuolana by some in order to distinguish it from the Portland cement or other types. This type of cement is of fairly good quality, approaching that of Portland cement, which it resembles; the color, however, being of a lighter shade.

On account of the scarcity of suitable gypsiferous deposits in many of the eastern states, slag cement is used to aid in supplying the breach caused by such deficit.

The immense deposits of gypsum in the west will preclude the manufacture of slag cements from that section of the country, for some time to come.

Lime.

Much of the lime used in New Mexico, up to a very recent date, was shipped into the territory from Texas, Colorado and Kansas; this, however, was not for lack of lime-making materials in the territory.

Lime, the principal constituent of ordinary plaster, is made by the burning or heating of limestone.

Any carbonate of lime, such as bones, marble, iceland spar, aragonite, shells of mollusks and chalk, when sufficiently burned, form lime.

Both the chemistry and technology of the process of lime making is simple and does not require the careful experience as in the manufacture of cement. Two things are requisite: the proper kind of limestone and that the same be burned sufficiently long at a high temperature. There is always more danger of under burning than over burning.

The object of burning is simply to drive off the carbon dioxide; the product left in the kiln is caustic lime.

The chemical equation, in presence of heat, is:

$$CaCO_3 = CaO + CO_2$$

which expresses the change.

The new product, caustic lime (CaO), has great affininity for water.

The chemistry of the milk of lime—the product run off into mortar boxes, when water is added to the quick or caustic lime—is shown in the following reaction with the evolution of heat:

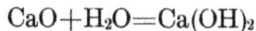

$$CaO + H_2O = Ca(OH)_2$$

The right hand member is milk or hydrate of lime or lime putty, and belongs to a group known in chemistry as *hydroxide*.

The blue or black limestones, which are very free from silica, usually make the best lime.

Mr. Charles Kempe of Tijeras, about twenty miles east of Albuquerque, is the largest producer of lime, in New Mexico. The capacity per week of his kilns, is about seven hundred

bushels.   The lime is made from a black limestone in the immediate vicinity, and is said to be ordinary in quality.
About two miles further up the canyon is another kiln, worked by some natives; the finished product is about the same as
that of Mr. Kempe's, but the output is not nearly so great.

A very fine quality of lime was formerly made at a point
called Fraley, about eight miles east of San Antonio, at the
northwest edge of the Carthage coal field.   These were the
largest kilns ever operated in New Mexico. When the branch
track of the A., T. & S. F. Railway was removed a few years
ago, operations were suspended.

At Las Vegas, near the Hot Springs, some large kilns are
operated by Mr. Fritz.   The production is considerable and
the grade of lime is excellent. A good quality of lime is made
from two kilns at Farmington. In fact all the principal towns
of New Mexico such as Silver City, Raton, Roswell, Las Cruces and Alamogordo, usually burn their own supply of lime
in the immediate vicinity.

The mode of mixing and preparing ordinary lime mortar
and its general uses are so familiar to all that any remarks
covering those points would be superfluous.

CHAPTER XXXII.

CLAY, BRICK AND STONE.

Nature seems to have abundantly endowed New Mexico with all that could be desired in the economics of building materials and ornamental stones.

Nothing definite has yet been done toward determining the extent, qualities and localities of these materials. The time for their full exploitation is still a matter for the future to determine. Present and past demands for such material have usually been supplied in the immediate vicinity where such needs have arisen. It is evident from the most cursory examination of the various fields, which have, as yet, been but little explored, that this Territory is amply provided with everything required in the line of the raw building materials.

Clay and Brick.

Very little concerning the clays of New Mexico is known. No detailed analyses of the various deposits have ever been made and our present knowledge of the character and extent of the beds is only fragmentary.

Clay may be defined as the silicate of aluminum and most always contains water in appreciable quantities.

Its origin is due to the decomposition of feldspathic rocks from which the oxides of potash and soda have been generally removed by meteoric and other agencies.

Kaolinite is the purest type of clay from which porcelain and the finest chinaware is made. The best grade of kaolin comes from near Janchau Fu, China. Germany and some other countries have almost as good an article as that found in China. The different types of clay are very numerous; passing as they do by insensible gradations from one to another, it is very difficult to draw any sharp line of demarkation, relative to their classification, only in a general way.

Chemical analyses and practical tests as to their commercial qualities, afford the chief basis in determining their fitness for use.

That splendid clays, necessary in the ceramic arts, for

bricks and other purposes, exist in the Territory, no one questions.

The aborigines were the first to utilize the clay products of New Mexico, as evidenced by their many peculiar wares in water jars, vases and baked images. Their beautiful coiled and decorated pottery have been much admired and commented on by archaeologists and curio collectors.

The more important uses of New Mexico clays, aside from pottery, when viewed from a commercial standpoint, would be in the manufacture of the various classes of bricks, terracotta, tiles, and sewer-pipe. Of these articles enumerated, none are manufactured in the territory excepting brick.

The clay bearing formation of New Mexico may be classed into four general groups, viz:

(1) Carboniferous clays.

(2) Jura-Trias and Cretaceous clays.

(3) Tertiary clays.

(4) Loess and alluvial deposits.

From the foregoing division it is seen that the clay and shale formations suitable for manufacturing purposes, exist in all the principal geological divisions, from the Carboniferous up to the river loess and alluvial.

Under the first division, both above and below the carbonaceous shales, which correspond to the eastern coal measures are found shales and clays which could be utilized in the manufacture of terra-cotta and brick. On the east side of the river at Socorro, clays supposed to belong to this age, have been extensively used at the Socorro Brick plant, making a good fire brick, due to their refractory qualities. Similarly, in the Lemitar mountains, much material belonging to the Carboniferous was used at the same plant. So far as known, these two places embrace the only points in the territory where clays of this age have been utilized.

In the second division, many clays and shales of beautiful bright colors may be found throughout New Mexico. Of course care should be used to avoid any of the highly gypsiferous clays and shales of the red series. Above and below many of the coal veins of the Cretaceous, fire clay of very excellent quality exists.

This clay, which is associated with the coal beds has never been utilized, excepting only sparingly at Gallup.

At Las Vegas a splendid brick is made from the dark Cretaceous shales; this however, is not directly connected with the coal bearing series.

Nowhere in the territory have the Jura-Trias clays been touched.

Of the clays in the third division only one, to the writer's knowledge, has ever been worked; it lies near Escondido, north of Socorro. This deposit is a whitish-gray color and makes a good refractory brick. It was in this marl-bed, the writer picked up several fragments of the bones, and some fairly well preserved teeth of the Tertiary horse—presumably, *equus excelsus*. These fragments are now in the museum of the New Mexico School of Mines.

The bones of this primitive equine were excavated some twenty feet below the surface and were scattered promiscuously about by the Mexican workmen and only by chance were any of the crumbling remains saved.

Near the mouth of Blue canyon in Socorro mountain, beds of kaolin exist which have, also, been worked by the Socorro brick plant. The age of these deposits is in doubt; though the disturbances rendering possible the condition of the kaolinization of the andesite, must have taken place during the earliest Tertiary times.

Division four is the most important source, commercially, of New Mexico's clays; it is from the river loess and alluvial deposits that most all of the bricks of the territory are made. Since the greater portion of the population is confined to the river valleys, it is cheaper to take advantage of the river deposits in the manufacture of bricks, than have the material brought from a distance.

This would account for most of the bricks being made from these clays, since the element of quality is not such a consideration as cheapness in manufacture.

Were quality only to be considered a factor, valley bricks would be relegated to the rear, on account of their inferiority in strength.

Mr. J. S. Macgregor of the Department of Mechanical Engineering in the New Mexico College of Agriculture and Mechanic Arts at Mesilla Park, kindly furnished me with the data on the crushing strength and absorption power of bricks coming from the Mesilla valley, Albuquerque and Las Vegas:

TABLE 1.

ULTIMATE CRUSHING STRENGTH OF COMMON AND MACHINE MADE BRICK.

Kind of brick.	Inches thick.	Area exposed to crushing sq. in.	Ultimate load in pounds.	Crushing strength in pounds per sq. in.	Color.	First crack observed.	Size of brick.
Mesilla Valley brick (Las Cruces)	2¼	4x4⅛	8,000	484.8	Light red	6 500	2¼x4x8¼
" " " (College)	2½	4x4	12,700	793.7	" "	6,500	2½x4x8⅛
Las Vegas brick (1) machine made	2¼	4x4	49,340	3,083.7	Buff	37,900	2¼x4x8
" " " (2) " "	2¼	4x4	46,150	2,884.3	"	34,000	2¼x4x8
" " " (3) " "	2⅜	4x4	42,100	2,631.2	Salmon red	30,880	2⅜x4x8
" " " (4) hand made	2¼	3½x4	39,340	2,809.8	" "	23,750	2¼x3⅞x7¾
Albuquerque brick (1) hand made	2¼	4x4⅛	33,450	2,027.2	Reddish brown	13,950	2¼x4x8¼
" " (2) machine	2⅜	4x4¼	54,060	3,180.0	Dark brown	28,000	2⅜x4x8½
" " (3) "	2¼	4x4¼	*........	....	Buff	....	2¼x4x8½
" " (4) hand	2¼	4x4⅛	12,260	743.0	Salmon red	8,250	2¼x4x8¼
" " (5) "	2⅜	4x4	16,000	1,000 .0	Dark red	10,520	2⅜x4x8

* Beyond 60,000 pounds.

TABLE II.

ABSORPTION.

Kind of brick.	Weight Dry		Weight after 24 hours in water.		Absorption in ounces.	Absorption %
	lbs.	ozs.	lbs.	ozs.		
Mesilla Valley brick (Las Cruces)	2	4½	2	11½	7	16.1
Mesilla Valley brick (College)	2	12	3	2½	6½	12.9
Las Vegas brick (1)	2	7½	2	14	6½	14.3
" " " (2)	2	6	2	11½	5½	12.7
" " " (3)	2	5½	2	12	6½	14.8
" " " (4)	2	4½	2	8½	4	9.9
Albuquerque brick (1)	2	6	2	11½	5½	12.7
" " (2)	2	10	3	2	8	10.0
" " (3)	2	8½	3	0	7½	15.7
" " (4)	2	6	3	2	12	24.0
" " (5)	2	4	2	10	6	14.3

By an inspection of Table I, it is seen that the machine-made brick gives a much higher crushing strength than the hand-made; and the buff-colored bricks, at both Las Vegas and Albuquerque give the highest crushing tests of all.

Also, the absorption power of these two varieties does not vary a great deal in proportion to their respective densities. Hand-made bricks usually have the greatest absorption and consequently show the lowest crushing strength; in the tables, No. 4 of the Las Vegas bricks seem to be an exception to the latter statement.

The only paving brick made at present in New Mexico, is by convict labor at the territorial penitentiary, Santa Fe.

Much of this material is being laid in walks and paving which may be seen about the Capitol at Santa Fe, in the City of, Santa Fe, in Albuquerque, Las Vegas and other places. This vitrified brick is of superior quality and finish; the clay comes from a deposit not far from the penitentiary.

All of the principal towns of the territory engage in the manufacture of brick for their own use; the alluvial and loess clays which are near at hand naturally receive the greatest favor.

In order to meet the demand for a cheap strong building material, sand-lime brick will likely become a new feature in New Mexico soon. A company is now being formed in Albuquerque to erect such a plant at an early date. The sand on the river makes this enterprise possible, since it is convenient and exists in practically unlimited quantities. The process is steaming and compressing; silicate of lime furnishing the bond. This brick will withstand an enormous pressure, is very compact and durable. Coloring matter may be introduced to give various shades of color.

The New Mexico Stone Manufacturing Company of Albuquerque, began making cement brick or rather blocks in the spring of 1904, and is meeting with marked success.

The following are a few analyses of clays from various parts of the country which may be used for comparison, viz:

Kaolin (China clay) $3H_2O. 2Al_2O_3. 4SiO_2$.

(Dana's Mineralogy)

Silica............................... 48.0%
Alumina 41.2

Water................................ 10.8
                                        ─────
        Total........................ 100.00%

Kaolin from Red Mountain, Colorado:
                (Dana's Mineralogy)
        Silica........................ 45.57%
        Alumina....................... 41.52
        Water......................... 13.58
                                        ─────
        Total........................ 100.67%

Crucible clay, Denver Fire-Clay Co., from Golden, Colorado:
                (U. S. Geological Survey)
        Silica........................ 71.81%
        Alumina....................... 15.09
        Ferrous oxide................. 1.75
        Lime.......................... .14
        Magnesia...................... .05
        Alkalies...................... 1.02
        Water......................... 10.14
                                        ─────
        Total........................ 100.00%
    The total fluxes in this clay are 2.96%.

Cretaceous clay from near Capitan, New Mexico:
        (Analysis by Hughes & Crickett, El Paso, Tex.)
        Silica........................ 58.4%
        Aluminum...................... 18.7
        Iron Oxide.................... 2.5
        Lime.......................... 6.0
        Magnesia...................... trace
        Water......................... 10.2
                                        ─────
        Total........................ 95.8%
    The deficit is, perhaps, all carbon dioxide.

East Slope of Sandia Mountains, Cretaceous Clay:
        (Analyst, F. C. Lincoln, N. M. School of Mines.)
        Silica........................ 65.54%
        Alumina....................... 25.76
        Lime.......................... trace
        Magnesia...................... trace
        Water......................... 6.70
        Organic Matter................ 1.36
        Undetermined (alkalies, etc.). .64
                                        ─────
        Total........................ 100.00%
    This is strictly a refractory clay, since it is free from any fluxes,
whatever.

Acequia Clay, Old Albuquerque:
        (University of New Mexico, D. W. Johnson, Analyst.)
        Silica, free.................. 48.33%
        Silica, combined.............. 28.21

```
Alumina................................   5.90
Iron oxide ...........................   3.41
Magnesia..............................   trace
Lime..................................   1.54
Water combined........ .. . .........   2.00
Moisture...... ...........  .....   4.28
Carbon dioxide............. ........   3.16
Potassa............... ............   2.43
Soda...........,..... .... ........    .80
                                       ────────
       Total ...................  ......   100.00%
```

According to the following figures given by the U. S. Geological Survey, it is seen that the clay working industries of this country are assuming enormous proportions and rapidly increasing from year to year.

The product reported for 1901 was valued at $110,211,587, increasing to $122,169,531 in 1902; making a gain of $11,957,-944 or 10.85 per cent.

During the year 1900, the clay product, which was made into brick, etc., in New Mexico, was valued at $26,900; this increased to $81,345 in 1901.

Building and Ornamental Stones.

Our knowledge of the building and ornamental stones of New Mexico is as limited and meager as that of the clays.

Until within the past few years no special want was felt to exploit the fields in this particular line. As previously stated, the material closest at hand was the kind utilized by the inhabitants on the frontier, who generally occupied the land adjacent and bordering on the principal streams.

Since suitable building stones were never quite so convenient for immediate uses as the adobe dirt, the latter has always been looked upon as a favorite kind of building material, since it possesses the elements of both cheapness and durability.

Within a radius of twenty-five miles the towns of the territory may be supplied with most kinds of building stones.

Since the compound parts of rocks are minerals, a rock properly defined would be a mineral aggregate. Rocks are classified with respect to the origin.

Three general classifications are recognized; igneous, metamorphic and sedimentary. Under these three divisions, the rocks of economic importance found in different parts of New Mexico, may be tabulated as follows:

Igneous:	*Metamorphic:*	*Sedimentary:*
Granite	Gneiss	Sandstone
Trachyte	Serpentine	Limestone
Rhyolite.	Quartzite.	Dolomite.

Igneous rocks are the types which have solidified from a molten magma; these are termed the crystalline series.

Metamorphic rocks embrace both the igneous and sedimentary types, which have been radically altered by dynamical agencies; these are termed the metamorphic series.

Sedimentary rocks are those which have been laid down in bedded planes due to the action of water; these are termed the sedimentary series.

Since the processes of nature are continuous, we are not wholly certain that any part of the original crust of the earth exists as it solidified from the Plutonic magma; but on the contrary it may have been regenerated a number of times.

The above classification may be regarded as general, for many varieties exist under each special type passing by tentative gradations, into other existing types of the series.

The value of a stone for building purposes depends largely on two physical conditions: (1) strength, (2) durability.

The tests of strength are three in number, viz:

1.—Crushing strength.

2.—Transverse strength (modulus of rupture).

3.—Modulus or coefficient of elasticity.

In order to determine the modulus of rupture:

Let,

W=concentrated load at center in lbs.

b=breadth in inches

d=depth in inches

l=length

R=modulus of rupture in lbs. per sq. inch; then from a principle in statics

$$W = \frac{2bd^2}{3l}R$$

$$\text{whence,} \quad R = \frac{3l}{2bd^2}W.$$

TABLE III.

CRUSHING STRENGTH OF BUILDING STONE.

Class	Size of piece tested in inches.	Color	Ultimate load in pounds	Crushing strength lbs. per sq. in.	First crack observed	Remarks
Sandstone........	2x2	Red	12,900	3,225	11,322	From eastern New Mexico. Cementing bond a lime carbonate.
Sandstone........	2½x3	White	32,210	4,295	25,650	From Albuquerque (Sandia Mountains). Silica the cementing bond.
Sandstone........	2½x3	Buff	52,780	7,037	36,470	From Albuquerque (Sandia mountains). Bond, iron oxide and magnesium carbonate.
Tecolote dimension..	4x4	Brown	*	....	....	From Las Vegas (Tecolote). Bond. Silica and lime carbonate.
Sandstone........	2½x3½	White	58,086	6,636	40,000	From Las Vegas, used for rubble work. Bond, the silicate of lime.
Sandstone........	4x4	Cream Yellow	50,600	3,162	44,000	From Las Vegas, footing stone. Cementing bond the oxide of iron and silica
Sandstone........	2½x2½	Brown	32,680	5,288	22,695	From Las Vegas, dimension stone. Cementing bond of silica.
Tufa or lava stone...	3x3	Buff	40,250	4,472	29,230	Lava from N. W. of Las Cruces. Cementing bond alumina.
Limestone (Imp.).....	2½x2½	Slate	24,600	3,936	13,200	From mountains east of Las Cruces. An imperfect limestone.
Marble........	2½x2½	White	6,750	1,080	Indefinite	From plains west of Las Cruces. Crystalline structure, fine.
Marble........	3x3	Bluish White	8,000	889	Indefinite	From plains west of Las Cruces. Crystalline structure, coarse.

*Beyond 60,000 lbs.

The tests of durability are four in number, viz:

1.—Specific gravity (weight of stone per cubic foot).
2.—Porosity.
3.—Temperature (extreme heat, freezing and thawing).
4.—Effect of acids (carbonic and sulphurous).

The first and second of this latter classification are not so important as those of three and four.

Under the latter two divisions would come the test of weathering; this, by far, is usually a more important factor to look into than any other test to which a stone may be subjected. If a stone will not stand weathering it is of little use as a building material or in macadamizing thoroughfares.

Beauty in color of building stone is always desired though it is not requisite to either strength or durability.

Some stones are so constituted as to be easily affected by atmospheric agencies; especially from carbonic acid which is present in the air to a small degree.

It is oftentimes observed in stone structures that some colored compound oozes out of the pores of the rock and stains the whole building. This stain is usually white or of a reddish-yellow color, and is due to the presence of alkalies or to the oxidation of small particles of iron pyrites, which exist in the stone.

East of Albuquerque in the Sandia mountains some splendid quarries of granite, sandstone and limestone are opened up. The limestone is obtained in Soda Springs canyon.

The granite is of coarse texture, containing some very large crystals of feldspar and hornblende, and of a gray color; it is quarried in the Carnuel canyon. A drab-colored arkose sandstone is also used extensively in Albuquerque; it is a splendid building stone and comes from Soda Springs canyon. A white fine grained sandstone comes from the Tijeras section and is very popular for window and door sills and footing stones of brick columns.

West of Las Cruces is a mottled marble of a very good quality; the crystallization is both fine and coarse grained.

Southeast of Lordsburg about twelve miles is a rhyolite rock of nearly a white color and is a most valuable building stone.

Silver City uses limestone and a dark colored curly marble

found in the vicinity. Other stones of an igneous origin are also used. Lordsburg is the nearest accessible railroad point to a valuable quarry of ricolite. No prettier ornamental stone is to be found in the west.

The quarry is on the Gila river north of Red Rock postoffice and is too far from transportation to be handled at a profit. The colors are yellow, blue, green and black of various shades and beautifully blended in a banded combination. It has a specific gravity of 2.57.

The analysis is:

Silica...............................	43.52%
Aluminum............................	16.88
Magnesia............................	23.78
Iron oxide.	trace
Lime...............................	2.22
Soda and Potassa.....	2.50
Water (combined)........	11.10
Total	100.00%

Some of this stone has been used in several of the larger buildings in Chicago, since it is susceptible of high polish.

Las Vegas is noted for its beautiful sandstone; for paving and dimension stones these quarries have no equal in New Mexico.

There are three varieties of these sandstones, the dark red, gray and brown. The paving stones are the dark red, while the building stones embrace all three colors. There are four principal quarries at Las Vegas. The New Mexico Normal is built of the dark red variety.

At Raton a splendid gray type of sandstone is the kind most used; this comes from three different quarries.

The building stone most used at Socorro is a light gray trachyte; it comes from the Socorro mountain. It is easily quarried and worked; the main structure of the School of Mines is composed of this stone.

Santa Fe county is not destitute, by any means, of good qualities of building stones. The beautiful cream colored sandstone used in the Capitol building came from a quarry on the hilltop at Lamy. Marble and good types of granite are found in the vicinity of Santa Fe.

In the vicinity of Roswell there is good sandstone and limestone used for building purposes in that city. The limestone is from quarries about five miles from town; the sandstone is some fifteen miles out.

North of Ancho about two miles is a small lead of a beautiful variegated flint; this stone takes a high polish and could be used to advantage in making small ornamental pieces. It is reported that a lithographic stone of most excellent quality exists in the Organ mining district.

At Gallup are some splendid Cretaceous sandstones which are utilized to good advantage in that town; a number of handsome structures having been constructed from that material.

Marble quarries near Alamogordo have recently been opened up with a view to supplying El Paso with this valuable product; the quality is said to be very fine, since the stone is of a mottled appearance; it is very attractive in ornamental work.

A good quality of marble is said to exist at White Oaks; the stone, as yet, has never found its way into the markets. White Oaks, also, possesses a beautiful brown type of sandstone, which has been used in the local trade to a certain extent.

Deposits of lithographic stone are reported in several localities; but little, however, has been done toward their development.

In the vicinity of Bluewater, south of the railroad in Valencia county, the New Mexico Pumice Stone Company is opening up what appears to be a first class lithographic material; the value of the product, of course, will depend on the size of the slab.

CHAPTER XXXIII.

MICA.

The mica deposits of New Mexico have, as yet, never received the attention to which they are entitled; chiefly, because the demands for that product have heretofore been supplied from localities nearer the manufacturing centers and from districts where transportation facilities were more favorable than in this territory.

By the increasing growth of the uses of mica in the varied industries, the time is near at hand when the New Mexico fields will be carefully exploited.

The greater quantity of mica mined in the United States comes from four states, viz: New Hampshire, North Carolina, Virginia and South Dakota. Production from the other states is comparatively insignificant. In the United States, the output of sheet mica was, during 1901, 360,060 lbs., valued at $98,859. Much mica is imported; the principal countries of export are Brazil, Canada and India. The mica from Canada is of a wine color and is said to be the best known for electrical purposes, owing to its great non-conductive properties.

Micas are, substantially, silicates of alumina; the varieties are determined by the presence of other material compounds.

Muscovite (white or transparent mica) is the potash variety; it is the kind used where transparency is essential.

Phlogopite (brownish-yellow mica) is the magnesium type; the kind most valued in electrical insulation. Biotite (dark or black mica) also, a magnesium compound; used for various purposes where economy is desirable.

The relative classes of mica, pertaining to market values, increase in geometrical progression:

	Relative weights.
Black and flawed mica	1
Smoked and discolored	2
White or transparent	4
Wine-colored, tough and hard	8

Muscovite is the most common of the micas, and is one of the principal constituents of many rocks. It is frequently derived from the alteration of some species of rocks; such as cyanite, feldspar, topaz, etc., and is thus of secondary origin.

The occurrence of the commercial forms of mica is associated with feldspathic-granite, pegmatite and felsite rocks.

Mica is infusible and is insoluble in acids. Clear, free-splitting mica is trimmed and assorted, usually, into four sizes, viz:

No. 1, measures 4x6 inches or 6x8 inches, and sometimes in larger pieces for special use; No. 2, averages 3x4 inches to 4x6 inches; No. 3, runs from 2x4 inches to 3x4 inches; and No. 4, averages from 1 inch up to 2x4 inches. No. 4 mica is known as rough trimmed mica. The waste or trimmings are sold at the mills for fine grinding, at from $15.00 to $20.00 per ton. A good quality of mica after it is mined will bring from $150 to $550 per ton.

With the sizes of the sheets, other things being equal, the value of mica increases in geometrical progression.

Insulation boards and bricks of mica called "micanite," are built up with the thin waste scraps, shellac being inserted between each flake, the whole is subjected to a 2,000 pound pressure per square inch. It is afterward steamed and remolded, then finally hardened in a kiln. The tremendous growth of electric appliances has caused an increased demand for insulating materials.

Ground mica is used in the manufacture of wall paper, giving it a spangled effect and also with the finer grades of paper a metallic white surface. As a lubricant for axle and journal bearings in heavy machinery it is of great value.

The first mention of mica in New Mexico was made by Lieutenant Pike in his Report of 1807. He says: "Near Santa Fe, in some mountains, a stratum of talc, which is so large and flexible as to render it capable of being subdivided into thin flakes, of which the greater portion of the houses in Santa Fe and all the villages to the north, have their window lights made."

This mica evidently came from the Cribbensville mines, near Petaca, Rio Arriba county; from Nambe, Santa Fe county; and from the little village Talco, in Mora county. It seems that the natives of New Mexico knew mica only as *talco*; hence, the name of the little village Talco, which is near the mica deposits in Mora county, as above mentioned. It also appears that these early people did not isolate the mineral *yeso*(gypsum) from the *talco*(mica). Since the selenite variety of gypsum occurs in divers localities in large transparent plates, it was used indiscriminately with mica, whenever transparencies were needed.

Down to a period of time as late as the American Occupation in 1846, there were no glass window lights in Santa Fe, excepting in the Old Palace. The most extensive deposits of mica found in the territory, so far as known at the present time, lie two and one-half miles southwest of Petaca in Rio Arriba county and are known as the Cribbensville deposits.

The geological occurrence is in feldspathic granite associated with white quartz and mica-schist.

Many of these claims are fairly well developed, especially the property owned by Moritz Leichtle. Considerable mica has been shipped from these deposits in recent years; at present the mines are idle. This property is ten miles west of the D. & R. G. Railway.

About three miles north of Ojo Caliente in Taos county, some fairly good indications of mica exist. One claim, especially, known as the Mica Age and belonging to Hon. Antonio Joseph, deserves mention.

The deposit occurs in a feldspathic-granite, very similar in character to that at Petaca. Some little mica was shipped from here about 1901. No extensive development has been done; the product shipped came from a sort of open cut and winze, not exceeding 25 feet in depth. This mica is, perhaps, a biotite and is not nearly so transparent as that at Petaca.

The mines at Nambe in Santa Fe county have been only meagerly exploited and the same will apply to the Talco deposits in Mora county. Several other localities have reported mica, yet no development has been done and the time is not quite at hand for the proper exploitation of this mineral product.

In the first canyon north of Mocking Bird spring, in the San Andreas mountains, southeastern Socorro county, some very nice books of mica float were found by the writer during a visit to that region in June, 1901; the float was not traced up to its source.

Good mica, it is said has recently been discovered in the Florida mountains, southeast of Deming; this rumor, however, has not been verified.

Sulphur.

Italy stands first in the production of Sulphur and has long been the principal source of the world's supply. The principal provinces of production are Caltanisetta and Girgenti, on the island of Sicily, in the region of Mt. Etna, where nearly one thousand mines are worked.

The occurrence of sulphur is generally associated with volcanic rocks, the cavities and interstices of which have become filled, due to the escaping of sulphuretted hydrogen gas. This probable reaction is best shown in the following chemical equation:

$$H_2S + O = H_2O + S.$$

The sulphuretted hydrogen on coming in contact with the air drops the atom of sulphur in exchange for its equivalent in oxygen; thus, water is formed and native sulphur produced.

In Italy the deposits occur in lenticular masses in veins and fault-fissures in what is supposed to be Miocene rocks.

Sulphur is found in most all volcanic regions associated with the eruptive rocks, occurring in the cavities, faults and on the shelving rocks in volcanic craters.

During the early Spanish conquests in the New World, sulphur was extracted from a number of places, and was used in the manufacture of gun powder. Cortez before laying siege to the City of Mexico, gathered a large quantity of sulphur which he used in making powder. This was taken from the shelving rocks of the crater of Popocatepetl, by lowering his men with ropes into the depths below.

For a similar purpose, Coronado and other early New Mexican explorers, utilized the sulphur from the Jemez Sulphur Springs from the old sulphur mines near Guadalupe and elsewhere in this territory. It is thus seen that sulphur was among the first minerals mined and utilized in New Mexico.

Not until the fall of 1901 did New Mexico produce and

refine sulphur on a commercial scale. The credit of this honor is due to Mariano S. Otero, who in that year completed a plant of five-ton capacity, and operated the same up to the time of his death, which occurred in April, 1904.

The deposits of sulphur at the Otero works are very considerable and are due to the eruptive and volcanic nature of the region.

It is said that the old mines of the sulphur deposits near Guadalupe are being looked into with a view of working them.

Near White Oaks and in the eastern part of New Mexico, contiguous to the Texas border, sulphur is found in the gypsum of those localities; the deposits have their duplicate in the extensive gypsum beds of Louisiana and western Texas.

Guano.

This product is used as a fertilizer and is always in good demand when the percentage of phosphoric acid runs sufficiently high.

The principal supply of the world comes from the small groups of islands off the western coast of South America. Guano is the accumulation of the deposit of birds found in places which they have occupied for roosting quarters for untold centuries.

In New Mexico this deposit is due to bats, and is found in caverns and shelving rocks of the vesicular lavas and along the limestone escarpment or rim-rock of the various mountain ranges.

The most important deposit of guano yet found in this Territory was in a volcanic crater, northeast of Engle, in the plains of the Jornada del Muerto.

This deposit was under the shelving rock in the crater; much of the rock and debris had fallen so as to almost completely cover the deposit. There must have been 3,000 tons of this deposit shipped to California during the years 1901-2. The lower or bottom part of this guano, resembled and was indurated as a rock.

Embedded in this deposit a mummy of one of the prehistoric people was uncovered, which had been wrapped in a cloth of coarse texture and shelved in this awful sepulcher centuries ago.

In Grant and Luna counties smaller deposits have been discovered and shipped at a considerable profit. Other deposits in the Black range, on the upper Gila and in the Caballo mountains are said to exist, but have never been fully exploited.

Pumice Stone and Tripolite.

There seems to be several localities in New Mexico where pumice stone exists.

The New Mexico Pumice Stone Company, promoted by E. B. Christy, architect of Albuquerque, has, perhaps, the most important deposit of this material in New Mexico. This deposit which is on the north side of the railway near Grant station in Valencia county, evidently owes its existence to the past volcanic action due to Mt. Taylor in that vicinity. The powdered material is equal to the imported product of the Old World.

On the opposite side of the Rio Grande from Socorro, about fifteen miles, a large bed of pumice stone occurs; the quality is splendid. This latter deposit is locally known as the "tripoli" beds.

Various other localities are reported to have a similar product, but no special investigations have ever been made of any of them, since no commercial importance has ever been attached to such material.

Tripolite is reported in a few localities, but it is to be doubted whether such a product has been found in New Mexico; usually the average person makes no distinction between these two products, yet they are formed under entirely different conditions.

Ocher (Mineral Paint).

In numerous localities throughout the territory are found deposits of ocher which cannot be excelled in quality for the manufacture of paints. Most every county has deposits of greater or less extent and quality. One of the purest beds of this class of minerals exists in the foot hills of the Sandia mountains, east of Albuquerque, in the immediate vicinity of Coyote springs. This deposit lies in a bed several feet in thickness and some portions of the strata would scarcely need grinding, the material being so fine. The colors are a beautiful red and yellow. This deposit is owned by C. C. and Brad Jones.

In the vicinity of San Pedro are large deposits of ocher which partake of most every tint imaginable.

The discussion of the mineral paints and the localities in which they occur would require a volume in itself to do justice to this natural product.

Alum.

New Mexico is said to have one of the most extensive alum deposits in the world.

These immense beds are on the upper Gila River about fifty miles above the L. C. ranch in Grant county.

This field embraces over two thousand acres of nearly pure Alunogen (hydrous aluminium sulphate).

After entering the narrow part of the canyon of the Gila River, deposits of alum are found at intervals in going up the stream for a distance of nearly forty miles, which terminate in the main beds some ten miles below the Gila hot springs.

These beds occur in the metamorphic and eruptive rocks of the region and their genesis is evidently due to solfataric action in the vicinity of the hot springs.

The mining district is known as "Alumina," and about 1,400 acres of the deposit were patented in 1894 by Lucien C. Warner and associates of New York. Lack of transportation prevents the product reaching the market; the nearest railroad point is Silver City, which is ninety miles distant.

Another large area of alum beds almost as extensive as the Gila River deposits exists in eastern Mora county, about twenty-five miles from Wagon Mound.

This deposit is not nearly so pure as that of the Gila River; an analysis made by the writer in 1902, gives the following results, to wit:

Silica	21.04 %
Alumina	7.23
Calcium oxide	.36
Magnesium oxide	.18
Sulphur	11.92
Water and volatile matter	59.27
Total	100.00 %

F. A. Jones, Analyst, May, 1902.

In the northwest part of Sandoval county, lying between sand rock in sort of a trough, is a deposit varying in width from fifteen to twenty feet and one thousand feet in length. Oxide of iron gives the deposit a reddish color.

Southeast of Springer on the ranch of Horace C. Abbott are deposits similar to those in Mora county.

The occurrence of the three last deposits is probably due to the oxidation of pyrites in bituminous shales and clays.

In the Sangre de Cristo mountains west of Red River post-office an extensive bed is said to exist; this report, however, is not verified. Smaller deposits, having no present commercial value, occur in numerous localities throughout the territory.

CHAPTER XXXIV.

GEMS AND PRECIOUS STONES.

Beside the fame of her turquoise New Mexico has in many instances produced some remarkable gems and precious stones.

The most plentiful gem is the peridot, found in the gravel washes on the Zuñi Indian reservation and at other points in the western part of the territory. These gems of the chrysolite are pronounced by experts to be the finest in the world. In the same locality and in Taos and Santa Fe counties are found many beautiful garnets; these gems occur in the gravels and are more or less associated with the peridot. A wide range in the variation of color is displayed in New Mexican garnets, that vary from a light rose to a bright red; this gem is frequently termed "ruby-garnet".

A few valuable emeralds or beryls have been picked up from the gravels near Santa Fe and are highly prized for their great beauty.

Some very good fire opals have come from the Cochiti country and also, from the vicinity of Santa Rita.

On a few occasions small sapphires and even diamonds have been accidentally found in gravel beds of certain localities in Santa Fe county; their occurrence, however, is very rare.

Euclase has been reported found in the vicinity of Taos. Agates, amethysts, tourmaline, quartz crystals, carnelian, moonstone and chalcedony are more or less common in the mountainous regions of New Mexico. It is not probable that New Mexico will ever take first rank as a producer of all the gems and precious stones; notwithstanding, it is likely to always remain in the lead in the production of turquoise, and hold a high position in peridots and perhaps garnets.

Turquoise.

"To wear a turquoise stone of blue
Will bring good luck and fortune, too."

The name *turquoise* is apparently from the French, which has its equivalent in Turkish; signifying that the gem first makes its appearance in Europe by way of Turkey.

The celebrated oriental gem comes from a district in Persia, near Nishapur, Khorassan, on the southern slopes of Mount Ali-Mirsa, northwest of the village of Maden.   An inferior turquoise comes from Mount Sinai in the Megara valley.   A greenish blue variety is found in the Kirgeshi Steppes, Siberia.

There are a few other localities of minor importance of the Old World from whence this gem comes.   An earthy variety comes from San Lorenzo, Chili, South America.

In the United States the gem is found in Colorado on the slopes of the mountain of the Holy Cross; in Nevada, about five miles north of Columbus; in Fresno county, California; in Alabama; in Arizona; and in New Mexico.

Of the various parts of the world in which this gem is found, it is now generally conceded that the purest stones come from New Mexico.

The production of turquoise in the United States for the year 1902 was valued at $130,000, against $118,000 for 1901; the greater portion of these values being accredited to New Mexico.

Odontolite, a fossil-bone or tooth, colored by a phosphate of iron, was used a few centuries ago, and is still used to a certain extent for jewelry; it is frequently passed for genuine turquoise.

Pliny speaks of the mineral *callais*, which is now generally regarded as turquoise.

There is but little doubt about the antiquity of turquoise in being used as an ornamental gem, in both the Old and the New Worlds.

The Mohammedans have used the polished gems for ornamental effects in their tabernacles of worship; as likewise did the ancient Peruvians and Mexicans in beautifying their temples of the Sun.   Who daresay, the throne of gold in the gilded palace of Tezones was not inlaid by the turquoise from the mines at Mount Chalchihuitl* close to Los Cerrillos and from the Burro mountains near Silver City?

Spanish or Indian work on the gold and silver mines of New

---

*Chalchihuitl, the Indian name for turquoise.   In the archives at Santa Fe is found the record of the Chalchihuitl grant bearing the date of 1763.

Mexico, prior to 1680, was very meager; and the evidence is scarcely sufficient to establish any pronounced degree of mining for the metals, excepting in a very few instances. But the mining of turquoise both at Los Cerrillos and in the Burro mountains, before the coming of the Spaniards, can scarcely be questioned. Immense excavations and old dumps, with which are associated relics of the stone age, practically verify the antiquity of those workings. Coiled pottery, the oldest known type, is in evidence; fragments of which are found in the old dumps both at Los Cerrillos and in the Burro mountains.

The writer with his own hands took from the old excavation at Mount Chalchihuitl, two stone hammers, which are here reproduced.

It is said that a stone hammer weighing some twenty pounds, with a portion of the handle still intact about the groove, was taken from these same excavations a few years ago. These stone hammers are made from a hornblende-andesite, common to the Cerrillos hills. The desiccated condition of the drift in which this latter relic was found would account for the preservation of the wooden handle.

Similar implements and tools of stone have been taken out of the old mines and dumps, in the Burro mountains, Hachita and in the Jarillas.

From personal observation, the writer is of the opinion that turquoise was hunted and mined by the aborigines of the west, many years and perhaps centuries, before the mining of the precious metals; and that mining for gold and silver was carried on but little in New Mexico by either Spaniard or Indian prior to 1800; with the possible exception of the old workings of *Mina del Tierra* and vicinity, near the turquoise mines at Los Cerrillos.

Perhaps much of the supposed turquoise used by the aboriginal Pueblo Indians, was malachite; they being unable to detect a difference between genuine turquoise and that mineral. It is quite common to find small pieces of the latter mineral in ancient graves and ruins, associated with pieces of the pure turquoise.

Turquoise, mineralogically defined, is a hydrous phosphate of aluminum, colored with a complex copper compound. The

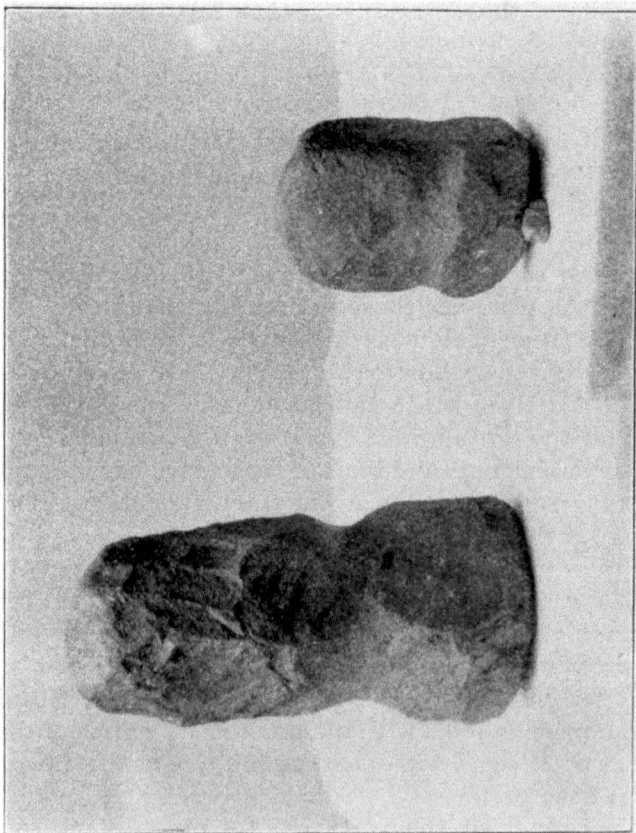

Fig. 43—ANCIENT STONE HAMMERS USED IN THE TURQUOISE MINES AT LOS CERRILLOS. Photographed by Florence E. Potter, July, 1904.

hardness of the stone is rated at 6; its specific gravity is from 2.62 to 2.89. It is noticed that the specific gravity of the turquoise of the western states is considerably greater than that of the Persian gem.

Some turquoise fades very rapidly a short time after being exposed to the air; this quality only retains its color so long as it is kept moist; while other qualities are appreciably affected by light. The various turquoise companies now have each a trade mark to protect the purchaser, should the gem not prove permanent in color or quality.

The trade mark is on the back of the gem; thus for the Gem Turquoise and Copper Company its trade mark is an ✕; for the Toltec Company a ⊤; for the American Turquoise Company an △: for the Azure Company a ○; and for the Himalaya Company an ←.

This guarantee of the genuineness of a stone in the protection of an inexperienced purchaser has had much to do with the popularity of the gem in the more recent years.

Professor Clarke of the United States Geological Survey gives the following analysis of a sample of bright blue New Mexico turquoise:

Aluminum and ferric oxide	39.53%
Phosphorus pentoxide	31.96
Copper monoxide	6.30
Lime	.13
Silica	1.15
Water	19.80
Total	98.87%

For comparison, an analysis of the Persian gem, taken from Dana's Mineralogy, is here given:

Alumina	40.19%
Phosphorus pentoxide	32.86
Copper monoxide	5.27
Ferric and ferrous oxides	2.21
Manganous oxide	.36
Water	19.34
Total	100 23%

Since the New Mexican and Persian gems are more highly prized than the turquoise from any other part of the world, it is remarkable how closely the stones agree in analysis;

coming as they do from diametrically opposite points of the globe.

Turquoise is found in four different districts of New Mexico, where it is profitably mined.

Taken in the order of their modern discovery the localities are: in the vicinity of Los Cerrillos, lying to the north of the railway, in Santa Fe county; the Burro mountains, southwest of Silver City in Grant county; at old Hachiti in Grant county; and in the Jarilla mountains, Otero county.

Much speculation as to the origin of turquoise has been indulged in by the professional geologist and others scientifically inclined. Since every turquoise field is somewhat different, no general theory yet propounded, would appear to account for its occurrence in the widely separated localities. Any hydrous aluminum phosphate would answer to the chemical composition of turquoise. The principal color constituent is due to a hydrous phosphate of copper, which chemically uniting with the hydrous aluminum phosphate would, under proper conditions, form turquoise.

From the chemistry of this compound, it would suggest the idea that we must look to its origin through aqueous rather than that of igneous agencies. Further it has been observed, in New Mexico especially, that the best specimens of the stone do not necessarily come from any great depth, but rather from beneath a superficial covering. At depth, apparently a thinning out of its occurrence seems to prevail or a greater increase in the percentage of copper is encountered, which renders the product worthless. Suffice it to say, that 100 feet in depth practically limits its occurrence as a paying marketable product. Again, the partially decomposed feldspathic rocks in which it is found in New Mexico, would cause us to look into the constituents or component elements which would be liberated therefrom, due to slow decomposition or kaolinization. It has also been observed that where turquoise exists, more or less copper is always found associated with the country rocks. Evidently, this kaolinization of the matrix or country rock would furnish the aluminum hydrate and phosphoric acid, the latter being derived from the apatite, which is found under the microscope, to be present in the Los Cerrillos trachyte, would combine with the hydrous phosphate of copper, thereby producing turquoise; all other conditions being favorable.

The finding of turquoise along certain thin seams, planes of cleavage or occupying cavities previously made in the country rock and by taking into consideration its reniform, stalactitic and incrusting deposits, would lead us to believe that its occurrence was due to deposition by descending surface waters, which carry in solution all the necessary constituents required in such a compound body. Some of the phosphoric acid may have also been derived from organic matter at the surface.

Thus, it is easy to account how the different grades, shades and qualities of turquoise may vary from that of the genuine type, when considering the varying constituents of the matrix or country rock in which the gem is embedded.

In all the districts of New Mexico where the gem is found, partial metasomatic action has taken place, due the deterioration through kaolinization of the feldspathic rocks.

This theory on the origin of the turquoise of New Mexico seems consistent with the nature of its occurrence and also agrees with what has been said concerning the gem at depth.

The antiquity of the Los Cerrillos turquoise has been briefly discussed in a preceding paragraph. Mount Chalchihuitl which lies to the north of the railway station at that point some three miles, is where the most extensive prehistoric and Spanish work was done.

In the elaborate ancient ramifications of the old workings at Mount Chalchihuitl, which were extensively prospected a few years ago, many stone hammers, whole vessels of ancient pottery and various crude mining implements were found. It is said that some twenty Indians were killed, about 1680, by the caving of a large portion of the works; this was claimed to be one of the chief causes which led to the general uprising of the Pueblos, that shortly afterward drove the Spaniards from the country.

Apparently the aborigines and early Spaniards exhausted this particular place of marketable turquoise; since considerable development was done a few years ago without any success. This hill or mountain is of a white or yellowish appearance and is different from the surrounding hills; decomposition by kaolinization seems well advanced. Whether this alteration has been hastened by escaping heat vapors or is due solely to surface and atmospheric agencies, it is somewhat difficult to conjecture; the former action seems most probable. The numerous intrusive dikes which traverse the district have no doubt played an active part in the general metamorphism of the associated rocks.

Bluish-green stains and streaks traverse this kaolinized rock in various irregular courses; it is along such lines of fracture that the marketable turquoise is likely to be encountered. Small seamlets and concretionary nodules, encased by the white or yellowish decomposed matrix are likely to contain valuable gems. Although, several tons of rock may frequently be broken and yet no valuable stones be found.

Some three miles to the northeast of Mount Chalchihuitl will be found the old Castilian mine formerly worked by the Spaniards. About the year 1885 the property was exploited and located by a man bearing the name of Palmerly. The Muñiz claim, one of the most important locations in the district was made in 1889, by F. Muñiz, a Mexican.

In 1891, C. J. Storey located the Sky Blue, Morning Star and Gem claims.

These latter five claims were bought by the American Turquoise Company of New Jersey about 1892, and are at Turquesa.

Also, near and adjoining the properties of the American Turquoise Company, J. P. McNulty has three locations which were made since 1892. Mr. McNulty has been the mine manager for the Tiffany people for a number of years. There are a number of other properties in the district which have produced beautiful gems; among which may be mentioned the Blue Bell and Consul Mahoney.

Burro District.

Much concerning the antiquity of the Los Cerrillos turquoise mines would apply to the mines in the Burro mountains.

The evidence, of very ancient work at these mines, may be seen in stone implements and fragments of coiled pottery found in the old dumps and shafts. The writer was shown a

sharp piece of iron, very crude and with an eye at one end
where a handle had been fitted, evidently a pick; which was
found in an old pit and must have been of Spanish origin.

The geology of the turquoise of this district is almost a
counter part of that of Los Cerrillos. A decomposed felspathic
granite, resembling aplite or Alaskanite and much kaolinized,
is the nature of the rock, in which the gem is found.

Fig. 44—"TURQUOISE JOHN."

The turquoise here is much the same as that at Cerrillos
and would take an expert to note any difference that might
exist.

An American by the name of John E. Coleman, "Turquoise
John," was the modern discoverer of turquoise in the Burro
mountains.

This gentleman located several claims in the later seventies
and early eighties. Nicholas C. Rascom, M. W. Porterfield

and T. S. Parker were also, pioneers in this section of the country and who became interested in a number of claims.

"Turquoise John," in 1882 disposed of his holdings to Messrs. Porterfield and Parker. These two gentlemen afterwards organized the Occidental and Oriental Turquoise Mining

Fig. 45—GEM TURQUOISE MINE.

Company; considerable development was done during the next several years. In 1901, a reorganization took place and the Gem Turquoise and Copper Company absorbed the holdings of the first company.

Since the organization of the new company much development has been done and this company now owns some of the most valuable mines of turquoise in the world.

There are a number of other properties in this district, which have good showings and may become important factors in the future production of this beautiful gem.

Jarilla District.

About 1898 some very valuable turquoise mines were exploited and opened up in the Jarilla mountains.

This district is accessible to transportation, since it is tapped by the El Paso and Northeastern Railway.

The principal group of turquoise claims here is known as the De Meules property. It was this gentleman and his associates who first opened up the turquoise here and made the district famous.

Some magnificent gems were taken out of the ground at a depth not exceeding forty feet, which proved a bonanza for the fortunate owners.

A number of other claims have been located and partially developed; but nothing has been found so valuable as what De Meules and associates discovered.

Eureka (Old Hachita) District.

Some turquoise of very fine quality has come from this district in the past years. There are a number of claims located and several have produced some very fine stones.

Ancient workings in one or two places exist and the turquoise mines here are in all probability as old as those of the Los Cerrillos and Burro mountain districts.

CHAPTER XXXV.

PETROLEUM AND ASPHALTUM (Bitumen).

Less than fifty years ago the value of petroleum was recognized only for its medical properties.

The year 1859 is generally recognized as the birth of the oil industry; a few barrels only were refined for use in the experimental trade of the country at that time. In 1902 the production of crude petroleum in the United States reached the enormous figures of 88,277,310 barrels; and the output of the world for that year aggregated 185,151,089 barrels. The United States takes the lead of all countries in the production of oil with Ohio the foremost state of the Union, and with Kansas a close second.

In the wild stampede for gold on the Pacific slope in '49 the prospector from the east passed over hidden treasures more wonderful and far greater in value, than any gold mines ever found. The existence of petroleum has been known from the earliest times and is alluded to in the Bible, in Chapter 1, Book 2, of the Meccabees. This Biblical allusion of the concealed fires of the priests was sought out two generations later at the time of the prophet Nehemiah, and found to be oil, which would burst into flames when poured on the hot sacrifical stones, used at that time in religious ceremonies.

Herodotus, 450 years B. C., speaks of a small river which empties into the Euphrates, as carrying down pieces of bitumen (asphalt) which was collected and used in Babylon in the construction of buildings. Similar statements referring to bitumen were made by Diodorus, Plutarchus and Quintus Curtius. Bricks and slabs of marble and alabaster and the grand Mosaic pavements of the ancient cities of Ninevah and Babylon were held together by semi-fluid bitumen.

The early Egyptians recognized the uses of asphaltum in many ways, such as in the calking and manufacturing of vessels and especially in bandaging the dead preparatory to burial.

Many interesting examples of the uses of petroleum and its resulting products by the ancients could be cited, but lack of space forbids their enumeration in this abbreviated volume.

Petroleum, in a general sense, embraces a series of hydrocarbons limited in one direction by a gas and in the other by a solid. It, therefore, is seen that hydrocarbons exist under three different conditions, viz:

1. Gaseous (where hydrogen equivalents exceed or equal those of carbon.)
2. Liquid (where hydrogen equivalents are less than those of carbon.)
3. Solid (where hydrogen equivalents are still further diminished.)

Petroleum in the United States is classified into two series, having different bases as follows:

1. Paraffine series, having a paraffine base.
2. Olefine series, having an asphaltum base.

Russian petroleum consists principally of the naphthene series.

In Pennsylvania, petroleum coming as it does from the older rocks, has a paraffine base; while in the west, in the more recent formations, it usually has an asphaltum base.

Indications are that the petroleums of New Mexico will generally have an asphaltum base.

The occurrence of petroleum does not appear to be confined to any particular geological horizon; since it is found from the lower Silurian, up to almost the latest geological epoch. Petroleum in its nature is migratory upward, and it is questionable whether any true primary deposits exist. The genesis of petroleum appears to be by destructive distillation of vegetable tissues that are embodied in bituminous shales. If this be true, it could not have remained in the rocks from which it was formed; it therefore would necessarily migrate and would thus become a secondary deposit.

In the eastern part of the United States the oil bearing strata are found in the Palaeozoic rocks; but in the west,

California in particular, the horizon is in the more recent rocks of the Miocene epoch.

Indications of petroleum are found in many places of New Mexico; yet, no locality has ever produced oil in any commercial quantity.

During and just after the time of the great oil excitement at Beaumont, Texas, in 1900-2, considerable activity was manifested in prospecting for oil in New Mexico.

A number of companies were organized and wells sunk to various depths (some quite deep), but without success.

Indications of the occurrence of oil have been noted in the counties of Leonard Wood, Colfax, Union, McKinley, Eddy, Lincoln, Otero, San Juan, Socorro and possibly Luna counties.

Northeast of Gallup a few miles, is a strong seepage of oil, and a well was put down to a depth of 900 feet, without any results. A second well was sunk by another company, 400 feet deep, at a point a few miles southwest of Gallup, in McKinley county, which was abandoned as fruitless. Some work, during 1902, was done on the Pioneer wells, near Farmington, in San Juan county, with no results. The site selected for boring was near the place where natural gas had previously been struck in drilling for water. In Colfax county, southeast of Raton, on the Wilson ranch and the ranch of Thomas Burns, a hole at each of these places was bored to the depth of 2,650 feet and 1,535 feet, respectively, but each well was finally abandoned, without getting anything more than a strong odor of oil from the bitumenous shales. The names of the two companies were the Raton Oil & Gas Co., and the New Mexico Oil & Gas Co.

At each of the above ranches there are springs with seepages of petroleum and the locations seemed to be decidedly favorable. When work stopped at the deeper well the sandstone horizon of the Dakota series was about penetrated by the drill.

A number of oil companies were organized, which prospected more or less about Santa Rosa, in Leonard Wood county, during 1902.

Of these latter companies, The Newman Oil Company prospected near the Pecos river and reached a depth of more than

1,000 feet. At 700 feet a small oil stratum was penetrated which proved too small to give results.

The Missouri Oil & Asphaltum Company prospected on the Jordan Plains, about twenty miles north of Santa Rosa, and the Continental Oil & Fuel Company also did some work about Santa Rosa, but nothing favorable was encountered during the time of operation. On the Perea land grant, a few miles north of Santa Rosa, a splendid bituminous sandstone exists in apparently large quantities; also, near by on Canyoncito creek petroleum oozes out of the earth and the product is used by the natives and ranchers for axle grease. Deposits of fire clay have also been observed in this locality.

Hon. W. A. McIvers of Nogal, who represented the New Mexico Oil & Development Company, found indications of the existence of oil between the head of the Mal Pais and the Gran Quivira.

Many other places throughout the territory are known to have indications of petroleum, that have never been investigated by an expert.

While the efforts of prospectors, so far, have proved futile in developing oil in New Mexico, it by no means should entirely condemn the field.

Since petroleum is very migratory in its habits, having always an upward tendency in its movement, its source may be too deep seated to have been penetrated by the bore-holes in the prospecting performed.

On the other hand, may it not be probable that the petroleum which once originally existed, has long since been dissipated, due to the effect of the numerous eruptive dikes and laccolithic sheets to which this particular portion of the earth's crust has been subjected?

Graphite.

Deposits of this material occur in several places in New Mexico; the quality, usually, is rather poor and of a semicrystalline character.

The most extensive occurrence of graphite in the Territory, known at the present time, is in the vicinity of Raton; this deposit seems to possess some commercial value and is being worked in a small way. Through the metamorphic action of eruptive dikes and laccoliths, this particular seam, at one time coal, has been transformed into a crystalline variety of graphite. The stratum is irregular and much of it is lacking at certain intervals, which necessarily renders the vein pockety. This deposit was laid down and is contemporaneous with the extensive coal fields of Raton and southern Colorado.

At the east end of Tijeras canyon, east of Albuquerque, near Whitcomb's spring, quite a band of graphite, presumably Cretaceous is observed, which has been thought by some to be coal and a little development work has been done under that supposition. Work at the time of the present writing is being conducted by Albuquerque parties in this vicinity, and thirty sacks of the earthy product have been sent to the Dixon Black Lead Crucible Company of Jersey City for analytical and practical tests.

In the Cretaceous sandstone in the vicinity of Taos, a prospector began development of a band of graphite which he supposed to be coal.

Graphite is used for making black lead crucibles, the lead for pencils, black paint, electrotypes and is a component of the best of lubricants for heavy machinery.

CHAPTER XXXVI.

RADIUM.

Much, of late has been written and said concerning the newly discovered and wonderful element, *radium*.

Many erroneous and misleading ideas have crept into the press relative to its nature, occurrence and chemistry.

The writer, at various times, has received communications from different parts of the territory, announcing the discovery of radium or that a "radium mine" had been found.

Still other inquiries were to the effect of ascertaining how the "mineral" could be recognized or determined by the prospector.

That radio active substances exist in certain New Mexican ores, can scarcely be doubted; and for the benefit of the prospector and miner it would be well for him to understand that no direct or practical method, at present, can be outlined to guide him in the field in detecting such elements in ores. In a general way, all ores of uranium, especially pitch-blende, should command his attention in his search for radium bearing substances.

Radium may be detected by impressions made on a photographic plate; the time of exposure varying from a few hours to several days; owing to the quantity of radium present. During 1896 M. Henri Becquerel discovered a certain form of energy in potassium uranium sulphate, which has since borne the name of Becquerel rays, in honor of the discoverer.

In 1898, Monsieur and Madame Curie announced the discovery of polonium and radium, and during the same year C. C. Schmidt discovered radio-activity in the metal thorium and its salts.

M. Debierne, in 1899, found a third radio active substance in pitch-blende and gave it the name of actinium.

There seems to be three radiations given off by radium, and they have been called the *alpha*, *beta* and *gamma* rays, respectively. If a strong magnet be placed near a beam of radium which has been isolated by means of lead-screens, it is found that the *alpha* ray is slightly repelled, the *beta* ray is attracted and the *gamma* ray is not affected.

The *alpha* ray is believed to be identical with the annode ray which is obtained when an electric current is driven through a high vacuum tube. If this be the case the atoms are charged positively—driven with a velocity of 15,000 miles per second. The *beta* ray has been definitely proven to be the same as the cathode ray of the vacuum tube; its material particles are thus negatively charged. The particles composing this ray weigh the one one-thousandth of an atom of hydrogen and travel at the rate of 90,000 miles per second.

The *gamma* ray is analagous and seems identical to the X-ray, the true nature of which has not yet been determined.

As yet, radium in its pure state has never been separated from its associated compounds, excepting in very thin films by electrolysis, but is obtained either as a chloride or bromide.

Monsieur and Madame Curie obtained the element as a chloride with barium; afterward the chloride of radium was separated from the barium ehloride by the slow process of crystallization.

Anhydrous radium bromide was first obtained by a German chemist; this process is somewhat more advantageous than that of the Curies.

The third method is the electrolytic process, which has the advantage of being very much quicker than the purely analytic methods, and may entirely supersede them after the nature of the element is more thoroughly understood.

As a matter of comparison, it may be said that more gold exists in the sea water than radium in pitch-blende.

Since the discovery of radium up to April 1, 1904, there perhaps, has never been a pound of its different grades produced altogether. At present, ores of uranium are the kinds most generally in demand; since they are more abundant and usually contain a greater quantity of radium or possess a higher degree of radio-activity than most other ores.

Uraninite or pitch-blende is the ore from which radium was first obtained by the Curies.

Later research, however, reveals the fact that this subtile

element is associated with a very large number of ores, more or less familiar to the miner and prospector.   In fact the chief difficulty is not in the finding of radio-active bodies, but substances that are not radio active; since, above the absolute zero everything in nature seems to radiate.

It is not to be understood that radium has actually been found in all the minerals named below, but they have become recognized distinctly radio-active, due to their absorbent qualities of radial emissions.

    Uraninite (pitch-blende)............Uranium
    Carnotite........... ....Alunite
    Columbite ....Thorium (compounds)
    Torberite ....Monazite
    Autunite. ...   .Tantalite
    Fergusonite...... ....Euxenite
    Samarskite............ ....Orangeite
    Cleveite............. ....Polycrase
    Xenotime ....Aeschynite
    Sipilite............................Niobite

Many solid bodies of various descriptions become radio-active, when placed in a closed vessel with one of the radio-active salts of barium, or what would be still more effective, to immerse the body or substance in a solution of such salt.

Copper, wax, lead, platinum, water, glass, etc., will partake of the radial emissions by absorption, when brought into the presence of such radio-active salts.

This induced radiation is identical in character with that of the original salt.

If a current of electricity be passed through a hermetically sealed tube containing a gas under pressure, the same would become luminous should the pressure be sufficiently reduced; air at 33-millimeters pressure would become luminous.

Should a radio-active substance be held sufficiently near such tube, luminosity would begin at a higher pressure, which for air is 44 millimeters.   It is observed that the color of the light varies in the two cases.

The "miracle of radium" is its power to perpetually emit heat and light without sustaining any apparent loss in weight or energy.   Under the generally accepted theory of the conservation of energy, radium seems to be an exception to this law of physics.

It is interesting to note, as before stated, that the emission of caloric energy seems to be perpetual and according to Mme. Curie's calculation, at a rate of about 90 centigrade calories per gram per hour.   Such emission of heat at this rate, according to Lord Kelvin in a recent lecture before the science branch of the British association, should it continue for 10,000 hours, would be sufficient to raise the temperature of 900,000 grams of water through one degree centigrade. It, therefore, would seem utterly impossible for this energy to come from a single gram of radium in 10,000 hours, without being replenished from some outward source.

In view of this fact it is not improbable that any losses of energy suffered by radial emissions may be supplied by ethereal waves.

An examination into the atomic weights of the rarer elements reveals the fact that they are singularly high; at least, their atomic weights exceed those generally of the more common or plentiful elements.

The atomic weight of radium as determined by Mme. Curie is given at 225; the supposition is that this is too low. The behavior of this element is such as to give us reasons to infer that atoms must have and do reach a limit of stability. After this limit is reached, dissolution or atomic decay of the elemental substance takes place.

Recent experiments have partly confirmed this statement, wherein it has been shown that the emanations from radium decompose and pass into helium.   The tendency of nature is then to reduce elements of high atomic weights to those of lower atomic weights.   Similarly, complex molecules in the organic world tend to break up into simpler substances. The processes of life elaborate complex material bodies, but as soon as those processes cease there is resolution into simpler products accompanied by a loss of energy.   We ask the question, is this transformation a kindred process in the inorganic world, and are the elements of higher atomic weight forever slowly disintegrating in favor of the elements of lower atomic weight?   Should this be true, the dream of the alchemist may, after all, be realized.

In the great laboratory of nature, the atomic decay of uranium may produce platinum, palladium or gold; and these would break down into copper or iron, and lead would decompose into silver.

If the forces of nature could be reversed by artificial means, copper or silver might be transmuted into gold and elements of still higher atomic weights.

Who knows but the discovery of radium has led us into the vestibule of the infinite and demonstrated the unity of matter; perchance, it is the grasping of the secret key of nature which unlocks the mighty mysteries of the physical universe?

It is too early to even conjecture the possibility of the effect which radium may produce in the treatment of disease.

Dr. Rollin H. Stevens, of Detroit, Michigan, has succeeded in provising a method in the treatment of cancer, by which no danger from burns is experienced, as was formerly the case in the X-ray and earlier radium methods.

The method pursued by Dr. Stevens is to induce radio-active principles to water by placing in the vessel containing water, a sealed radium tube which imparts its rays to the fluid of absorption. Radio-active water is thus produced, having a most powerful effect. Injected into cancers this water stopped all pain within ten minutes time. Apparently this method of treatment is curing every case in which the water has been applied. In each instance under this treatment the cancer has been rapidly reduced and diminished constantly in size, resulting in a permanent cure.

The possibility of natural or mineral springs waters becoming charged with radio-active elements through absorption and otherwise, has been suggested under the chapter on mineral waters.

Radio-activity defined, means the number of times a piece of radium is radio-active stronger than the influence of a piece of uranium, that is to say a substance has a radio-activity of 500 when its radial properties are 500 times greater than uranium.

On March 10, 1904, an importer of radium states that this product has advanced in price $4,200,000 per pound during the last two days. The commercial rate last week was $8,400,000 now it is $12,600,000.

Notes on Some of the Rarer Metals.

Platinum has a specific gravity about the same as that of gold. It was valued at $18.50 per ounce, the market price, March 17, 1904. It is the most useful of the platinum group of metals, which embraces palladium, rhodium, ruthenium, iridium, osmium and platinum. The three first are of little importance commercially.

Iridium, worth more than $780 a pound, is the hardest metal known and is used to tip gold pens.

Palladium which has the smallest coefficient of dilatation, is used for the mounting of astronomical instruments. The standard meter of France is made of palladium. The pure metal costs $4.82 a pound.

Lithium, worth more than $1,100 a pound, is used only in medicine, its salts being valuable in rheumatic affections.

Tungsten, worth 30 cents a pound, is largely used in metallurgy and gives to steel, qualities similar to those imparted by molybdenum.

Molybdenum, worth $1.44 a pound, is used in metallurgy. Molybdenum steel possesses the rare quality of preserving its hardness even when heated to redness.

Selenium, which has the curious property of losing its resistance to the electric current under the influence of light, is used in the telelectroscope and is worth $22 a pound.

Uranium, worth $86 a pound, is used in the glass and porcelain industries. It has been found that uranium carbide is superior to nickel or tungsten in the manufacture of high grade steel.

Vanadium oxidizes in air with great difficulty, melts at 2,000 degrees and becomes red hot in hydrogen. Neither hydrochloric acid nor nitric acid attacks it. It costs $592 a pound and is used in coloring glass and in making indelible inks.

CHAPTER XXXVII.

MINERAL WATERS.

In dealing with the waters of New Mexico no attempt is made toward their proper classification.

"Mineral water" in its broadest sense is water of any kind found in nature; since water itself is a mineral and is never found absolutely pure in its natural state.

The dissolving power of water is very great and it is due to this property that all water in its natural state is more or less contaminated with the material substances through which it courses.

Dr. Charles Daubeney, in his Sixth Report of the British Association for the Advancement of Science says: "The term *mineral water* in its most extended sense comprises every modification existing in nature of that universally diffused fluid, whether considered with reference to its sensible properties or to its action upon life." Any waters containing a large amount of mineral matter or characterized by an unusual degree of heat, and having a certain therapeutic effect upon the animal economy are usually classed as "mineral waters."

It is not within the sphere of chemical analysis to explain in all cases the therapeutic effects of any particular water in the cure of certain diseases; for many of the springs which have acquired reputations for their medicinal virtues are not so highly mineralized as some of the ordinary potable waters.

The cause of this must be accounted for in some other way. It is not improbable that many of the famous mineral springs are charged to a certain degree with substances of radio-activity, which would account for the beneficial effect experienced in certain ailments. It would be interesting to have experiments conducted along this line with the waters of some of the more prominent springs. The glow of the ocean water at night observed in the rear of a moving vessel is, perhaps, not fully explained by contributing the phenomenon as wholly due to phosphorus.

It is the opinion of the writer that the medicinal virtues of the waters of many springs are due to the presence of some of the more subtle elements that have, hitherto, escaped the search of the chemist in the ordinary analytic methods of analysis.

This assertion is borne out by the fact that a few of the most noted mineral waters contain less solid matter than is found in some of the common springs and potable waters; notwithstanding the fact that the same observed elemental compounds exist in each. Owing to the comparatively recent geological structure of the rock system in New Mexico, most all of the waters of every description carry much matter in solution; inasmuch, the formations and soils have not been leached to the extent as that of the eastern and central parts of the United States. Therefore, the waters of the Territory may be called with propriety "volcanic waters;" this term would especially apply to the thermal springs.

Ojo Caliente (Hot Springs).

These springs take rank with any of the celebrated springs in the United States and, perhaps, in the world for their marvelous curative powers.

Their waters have long since been a balm for suffering humanity; tradition points to their visitation by the aboriginal tribes of the southwest long before the occupation of the country by the Spaniards. Some of the most ancient ruins of the southwest are found in the valley and on the contiguous mesas to these springs. A later and more extensive city of the Pueblos is in evidence from ruins found on the mesa on the west side of the valley about half a mile south of the springs.

This Pueblo seems to be the Phœnix which has arisen from the debris of a still older city; thus the history of those primitive people exists only in the fragmentary ruins that have defied the elements to the present time. It is claimed that Cabeza de Vaca, the discoverer of New Mexico, in 1534, visited and named these springs Ojo Caliente, which name they have ever since borne.

This is quite probable, inasmuch as this unfortunate Spani-

ard became noted among the aborigines, during his wander
ings, as a "healer of the sick."

Many superstitions and traditions exist in the neighboring
Indian tribes today, concerning the efficacy of these waters.
The habit of bringing their sick to these springs every year
is nothing more than the inculcated practice of their long time
custom of past centuries.   Brush houses are erected along
the valley where the sick remain until a cure is effected.

One of their methods of treatment of rheumatic and blood
disease, besides drinking the water and bathing, is as unique
as it is effective. A "sweat-house" of adobe mud is first erected
which has the appearance of a Dutch oven; fires are built
near by and rocks heated which are placed inside the sweat-
house.   Blankets are laid over these warm stones and the
patient, after taking some of the salts from the evaporated
water, is put in the sweat-house until he perspires freely.
This is repeated most every day until the patient is well.
Cabeza de Vaca carried the news of the virtues of this wonderful water to Mexico, and it was thought by many at that
time to be the "fountain of youth."   Many of the syphilitic
troops of the Coronado expedition were said to have been
cured at these springs.

Some fifty years after the expedition of Coronado, a terri-
ble blood disease spread among the Pueblo Indians with such
fury and fatality, that the Spanish priests erected a church
on the plaza, just east of the springs, in the year 1590, to
accommodate the crowds of sufferers who sought spiritual
comfort while trying the efficacy of the springs.   No doubt
this great affliction wrought on the simple Pueblos was due
to the Spanish soldiers who tainted the blood of the aborigines at the time of Coronado.   This ancient church is still
standing in a fairly good state of preservation.   As near as
can be learned these springs were first visited by Americans
in 1810.

The dissolving power of this water is very great; hence,
for stone in the bladder, gravel or any calcareous affections
it is highly recommended.   Good effects are produced in
disease of the joints and muscles from electric mud baths, by
applying mud poultices to the affected parts.

There are four of these springs in a very small compass,

Fig. 46—OJO CALIENTE (Hot Springs).

and each one is peculiarly adapted for the cure of particular diseases.

Physicians recommend the water especially for rheumatism, kidney, stomach and blood disorders. The temperature of these springs varies from 90° to 122° Fah.

The following are analyses of the waters as reported from the Division of Chemistry of the Department of the Interior:

Constituents	Spring* No. 1.	Spring* No. 2.
Sodium carbonate	196.95	184.29
Sodium and magnesium carbonates.	6.25	5.40
Lithium carbonate	.21	.16
Iron carbonate	trace	trace
Sodium sulphate	13.60	19.33
Potassium sulphate	5.17	5.34
Sodium chloride	38.03	39.78
Silica	trace	trace
Parts in 100,000. Totals....	260.21	254.30

*Oscar Loew, Analyst, 1875.

The third spring of Ojo Caliente is the largest, and is properly classed as a chalybeate spring, due to the presence of the large percentage of iron carbonate.

Its waters contain 1,686.84 grains of alkaline salts to the gallon, which is extremely high.

The authority of this analysis could not be learned, but is supposed to have been made by the U. S. Geological Survey:

Constituents.	Spring No. 3 Iron Spring.
Sodium carbonate	196.95
Iron carbonate	20.12
Arsenic*	10.08
Magnesium carbonate,	6.10
Silicic acid	4.10
Calcium carbonate	4.20
Sodium chloride	40.03
Lithium carbonate	1.22
Potassium sulphate	5.29
Parts in 100,000. Total	288.09

The state of oxidation is not designated.

The fourth and last spring of the group is the celebrated
Lithia spring, and known as the new spring; the following is
the analysis:

Constituents.	Spring No. 4* (New Spring)
Lithium chloride	20.9
Potassium chloride	59.9
Sodium chloride	305.5
Potassium Iodide	trace?
Sodium arsenate	trace
Sodium borate	5.4
Sodium nitrate	trace
Sodium sulphate	223.3
Sodium carbonate	1846.9
Ammonium carbonate	trace
Calcium phosphate	.3
Calcium fluoride	10.7
Calcium carbonate	43.0
Barium carbonate	trace
Strontium carbonate	2.4
Magnesium carbonate	33.2
Silica	60.2
Iron oxide	1.6
Alumina	.5
Carbonic oxide	775.6
Parts in 1,000,000. Total	3389.4

*F. W. Clark, Chief Chemist.

Traces of potassium oxide, sodium arsenote, sodium nitrate, ammo-
nium carbonate and barium carbonate.

The chemist also notes that no organic matter is present
and that the state of oxidation of the iron is unknown, since
the iron was found deposited as sediment in the bottles.

The writer's opinion concerning the state of the iron of the
two last springs, is that it exists as ferrous carbonate in the
earth; under released pressure, on reaching the surface, the
carbon dioxide escapes, rendering the iron ferrous oxide,
which immediately passes to the ferric state on coming in
contact with the air.

The combined flow of these springs will approximate 300,000
gallons in twenty-four hours.

Owing to the comparatively recent volcanic disturbances in
Taos county, geologically speaking, other thermal springs in
this region are known. The Wamsley hot springs are gain-

ing some reputation for their beneficial effects on the animal economy.

These springs occur in the deep gorge of the Rio Grande, just below the toll bridge on the road leading from Taos to Tres Piedras. This water is little more than luke-warm and very similar to the warm springs at Glen-woody camp, about eighteen miles below. It is observed that the flow from each of these springs is from the west and on that side of the river.

Faywood Hot Springs.

Prominent among the mineral springs of New Mexico and the southwest that have become noted for their curative properties, is the celebrated Faywood Hot Springs of eastern Grant county.

This spring flows from the top of a mound or truncated cone, which it has built up out of calcareous matter and silicious sinter. This conical shaped mound rises to a height of nearly forty feet above the surrounding plain. For this type of springs it is representative and compares favorably with some of those at Yellowstone Park and Iceland.

The reservoir or surface end of the geyser tube is about fifteen feet in diameter and is said to have been sounded to a depth of six hundred feet; the truth of the latter statement is much to be doubted.

In 1893, A. R. Graham installed a large pump and lowered the water in this caldron, in order to cement the sides and prevent leakage and thereby cause the water to be elevated so it would flow out of the top of the mound. The walls of the geyser tube were found to be quite uneven and jagged and on some of these projections were found many interesting archæological relics which had either fallen in by accident or on purpose.

Several stone hammers, flint and bone implements, copper spoons, and earthen vessels were found; and last but not least, the bones of human beings were taken out. The skull of one of the poor wretches who had either fallen or been forced into this boiling caldron, is in the Whitehill collection at the springs hotel, where also, may be seen a number of the other curios taken from this frightful vent.

On a visit to the springs by the writer, May 3, 1903, and in

Fig. 47—FAYWOOD HOT SPRINGS.

examining the pile of debris which was taken from the vent almost a complete lower jaw bone of a human with the full quota of teeth was found and a number of flint implements, belonging to the stone age.

The action of the hot alkali water had boiled out every trace of organic tissue from the bones and bleached them almost perfectly white. From the foregoing it is evident that this spring had been known by the aborigines and may have been used for inflicting the death penalty on the more unfortunate members of their tribe. From the nature of many of the articles found as well as the human bones, it would appear that the more plausible theory of their presence in the vent is that they had fallen in by accident, during visitations of the sick in search of health.

C. C. Conrad, the man in charge of the bath house, accounts for one of the skeletons, at least, in the following bit of frontier history: Conrad said while in Silver City several years ago he met an old soldier who exhibited a diary which he had kept during his service in the army dating back to 1850. On one occasion he was detailed as one of a cavalry squad from Fort Selden to warn an old Dutch settler and his wife, who were then living a short distance to the west of the hot spring, of a band of Apache Indians headed that way. The cavalry troops camped that night at the spring and were reinforced there by a party from Pinos Altos. About daylight, not aware of the presence of the soldiers, the band of savages swooped down from the adjacent hills, expecting to surprise the Dutch family. The suprise came the other way, for the cunning savages met with a warm reception by a volley from the soldiers. One Indian fell wounded near the spring; the others fled in disorder. A soldier near by picked up the wounded Indian and thrust him into the boiling caldron before any one could interfere. The soldier was court-martialed for this cruel offence, but was acquitted.

The water of Faywood hot spring is alkaline and has a temperature of 142° Fahrenheit; it has as high a temperature as any spring in New Mexico, with two or three exceptions. At one time this spring may have been a geyser and spouted intermittently similar in manner to those of Iceland and the Yellowstone Park. Its capacity per hour is 6,000 gallons.

An analysis made by Professor Arthur Goss, Chemist of the College of Agriculture and Mechanic Arts, Mesilla Park, New Mexico, shows one gallon of the water to contain the following solids:

Constituents.	Grains.
Lime (CaO)	4.50
Magnesia (MgO)	1.46
Soda (Na₂O)	11.17
Potassa (K₂O)	1.52
Iron and Alumina (Fe₂O₃, Al₂O₃)	1.28
Silica (SiO₂)	4.10
Sulphates (SO₃)	4.21
Carbonate (CO₂)	9.98
Chlorides (Cl)	1.77
	39.99
Less oxygen equivalent of Cl	.40
Total solids	39.59

The hypothetical combinations were not calculated.

This water is especially noted for stomach and kidney troubles. Rheumatism yields to the sweat baths; while in skin and blood diseases some remarkable cures are reported.

Las Vegas Hot Springs.

These springs are perhaps the most widely known of any in New Mexico; the place has been famous as a resort, even before the advent of the Santa Fe Railway.

There are forty of these springs in number, all of which are situated in a very circumscribed area.

Many archæological relics, such as stone-axes, pottery, flint and bone instruments have been found in the immediate vicinity of the springs and in connection with near by ancient ruins of a prehistoric people. It is, therefore, evident that the Europeans were not the first people to drink and bath in the waters from these springs.

Mud baths and poultices, for swollen joints, due to gout and rheumatic affections, from the black tenacious mud about the springs, have proven very efficacious.

The beautiful scenery and mountain air in connection with the water, always produce beneficial effects, provided the patient's vitality has not previously been exhausted, on reaching the springs.

Water from different springs varies from 75° to 140 Fahrenheit; the character is alkaline-saline.

The following are analyses of four of the springs as made in 1875 by Oscar Loew, for the Department of the Interior:

Constituents.	Hot Spring.	Spring No. 1.	Spring No. 2.	Spring No. 3.
Sodium carbonate	120 00	1.72	1.17	5.00
Calcium carbonate..... Magnesium carbonate.	} 13.75	9.08	10.63	11.41
Sodium sulphate..........	5.26	14.12	15.93	16.27
Sodium chloride.......	6.41	27.26	24.37	27.34
Silica	trace	1.04	trace	2.51
Potassa..............	....	trace	trace	trace
Lithia...............	....	strong trace	strong trace	strong trace
Parts in 100.000. Totals.	145 42	53.22	52.10	62.53

Jemez Hot Springs.

At the time of Coronado's expedition the Jemez country was a thriving, populous province. Many large pueblos then existed, which were visited by several of the captains of the Spanish army at that time. It seems that the prehistoric races knew that certain medicinal virtues existed in many of the volcanic waters, which accounts for the selection of convenient sites, near such waters, of a number of their villages.

There are two groups of these springs; they are in San Diego canyon, and are designated by the upper and lower groups. The upper group consists of forty springs; the temperature of the water ranges from 70° to 105° Fah. This group is about fourteen miles above Jemez, in Sandoval county.

The lower group embraces ten springs and the temperature varies from 94° to 168° Fah. The temperature of one of these springs is the highest in the territory, insofar as present information goes. The waters of both groups are saline :

Constituents.	Geyser.	Spring No. 3.	Spring in upper group.
Sodium carbonate......	......	......	.0219
Calcium carbonate.....	.0641	.0300	.0548
Magnesium carbonate...	.0103	.0240	.0057

Iron carbonate.........	......	.0002	......
Sodium sulphate........	.0035	......	.0059
Calcium sulphate.......	trace	.0262	.....
Sodium chloride........	.1622	.1508	.2642
Silica	trace	.0010	.0201
Potassa................	trace	trace	trace
Lithia..................	trace	trace	trace
Phosphoric acid	......	......	trace
Parts in 100. Totals....	.2401	.2322	.3726

San Ysidro Spring.

San Ysidro spring is near Jemez and is one of the results of the volcanic conditions in the Jemez country.

It is classed as one of the mineral waters of the territory and is considered to be important by many in its medicinal virtues.   The following analysis of this water was made by Oscar Loew in 1875, under the supervision of the Department of the Interior:

Calcium carbonate...................................	.0670
Magnesium carbonate...............................	.0243
Iron carbonate..............	.0008
Sodium sulphate..........	.1639
Sodium chloride....................................	.3072
Silica.............	trace
Patassa.....	trace
Lithia	trace
Parts in 100. Total.........................	.5632

The water of this spring is carbonated.

The Sulphurs.

These springs are located in Santiago canyon, which opens into the Jemez creek. They are two miles north of the Jemez springs at an altitude of 6,740 feet.   The temperature of their waters ranges from 70 to 105 degrees Fahrenheit.   They flow from caves of carbonate of lime, the caves being from a few inches to twenty feet in height.   Combined, these caves form a lime ridge 30 feet high and 200 feet long.   The springs carry .3726 parts of solid matter in 100 parts of water; the mineral carried being chloride of sodium, sulphate and car- bonate of soda, lime and magnesia and the waters resemble those of Marienbad.   The springs are both mud and vapor; being strongly impregnated with sulphur, they are much sought by persons suffering with rheumatic or syphilitic complaints.

San Antonio Springs.

These springs are four to six miles west of the sulphurs nd are also heavily impregnated with minerals giving them medical properties similar to those of the Jemez hot springs.

Aztec (Ojo Xigante) Spring.

About four miles east of Santa Fe is a spring, the waters of which are becoming favorably known and are being used to a consirable extent by the people of Santa Fe and elsewhere. This spring is known as the Axtec (Ojo Xigante); since, like other watering places, it was frequented by the aborigines.

While the solid matter contained in the water is not so great as that found in many other springs in New Mexico, nevertheless the water has beneficial effects in stomach and liver troubles As previously noted it is not absolutely necessary that the solid constituents should run extraordinarily high to be effective as a good mineral water.

F. W. Clarke of the U. S. Geological Survey, in 1885, gives the following analysis of the Aztec spring, which was made at the request of an army officer who had been drinking the water when stationed at Fort Marcy, and who first recognized its beneficial effects on himself and troops.

Calcium carbonate	.1538
Magnesium carbonate	.0605
Sodium sulphate	.0225
Calcium sulphate	.0050
Sodium chloride	.0193
Silica	.0220
Parts in 100,000 Total	.2831

In a foot note the chemist adds, that the water contains enough carbonic acid to retain the carbonates of calcium and magnesium in solution as bi-carbonates.

Coyote Canyon Springs.

These springs are situated in Coyote canyon, at the southwestern base of the Sandia mountains, about fourteen miles southeast of Albuquerque, in Bernalillo county.

Three different places are noted in this canyon for their water. The Harsch springs are the ones first met in going up

Fig. 48—COYOTE (CHAVEZ) SPRINGS, Near Albuquerque.

the canyon; the next are the original Coyote (Chavez) springs; and the third is Topham's (artesian well) spring. Originally, this water became known through the main Coyote (Chavez) springs, the principal ones in the canyon, from which all of the waters now are celebrated as "Coyote canyon water."

Afterward, some persons located and developed a mining claim known as the Cotton-tail mine, now the Harsch carbonate spring, sinking to a depth of about eighty feet. This shaft encountered a flow of water, which afterward filled and flowed out through an escape, a few feet below the top.

The soldiers of Coronado frequented the main Coyote (Chavez) springs as early as 1541–2 while stationed near Albuquerque; and Friar Ruiz and his co-laborers gave thanks to the Almighty on reaching this oasis of refreshment and rest.

All of the waters of Coyote canyon are especially beneficial for kidney and bladder troubles. All the waters of this canyon are surcharged with carbon dioxide; hence, they are classified as carbonated waters.

Approximately, 10,000 gallons of the Topham's Artesian Coyote water were bottled during 1903, and 25,000 gallons were marketed from the bottling plant of Harsch.

It is observed from the similarity and quality of the solid matter, as determined by analyses of these waters, that they, perhaps, come from the same source and are practically the same waters.

The analyses of the two Harsch springs made at the College of Agriculture and Mechanic Arts, at Mesilla Park, New Mexico, show:

Constituents.	Iron Spring.	Cotton-tail Spring.
Sodium chloride	67.60	53.34
Sodium sulphate	8.31	10.87
Sodium carbonate	12.18	4.24
Potassium sulphate	8.33	5.03
Calcium carbonate	57.15	54.76
Magnesium carbonate	16.11	15.79
Iron, silica and alumina	6.70	.66
Organic matter & water of crystallization.	7.92	3.01
Parts in 100,000. Totals	184.30	147.70

Analyses of the original Coyote (Chaves) spring and the Topham Artesian (well) spring, are as follows:

Constituents.	Chaves Spring*	Artesian (well)+
Iron carbonate..............	13.00	.4
Magnesium carbonate	13.40	13.8
Calcium carbonate............	48.60	45.7
Sodium sulphate	14.00	8.0
Sodium carbonate............	6.24	....
Sodium chloride...	58.60	19.1
Silica.......	trace	1.4
Grains per U. S. Gallon. Totals	153.84	88.4

*Analysed by Ed B. Jorgenson, chemist.
+John Weinzirl, chemist, 1901, University of New Mexico.

Both of these waters, also, contain traces of potassium and lithium salts, calcium sulphate, phosphates and free carbonic acid gas exists in quantity.

Whitcomb Springs.

In Tijeras canyon about eighteen miles to the east of Albuquerque, a pleasant and beautiful resort is found at Whitcomb's springs. The great tilted block-like mass which composes the Sandia mountains is quite attractive to visitors and the inhabitants of Albuquerque, during the heated season, and camp Whitcomb is by nature a favorite spot for the rendezvous of the crowds bent on rest, pleasure and sight-seeing.

The analysis of Whitcomb spring water, as made by R. W. Tinsley of the University of New Mexico is given in the following:

Sodium chloride	.1927
Calcium sulphate.....................	1.4360
Calcium carbonate......	8.1896
Magnesium carbonate..........................	1.5188
Total grains per U. S. gal.	11.3371

Las Palomas Hot Spring.

This spring is in Sierra county on the Rio Grande, and can be reached from Engle, the nearest station on the Atchison, Topeka and Santa Fe Railway.

The spring, though somewhat obscure when compared with some of the more widely known springs of New Mexico, stands preëminent in its curative qualities in cases of rheumatic affections.

The following is an analysis made, presumably, by the College of Agriculture and Mechanic Arts at Mesilla Park:

Constituents.	Parts in 100,000.
Lime.........	22.50
Magnesia..	3.10
Soda	100.32
Potash.	7.00
Silica................	4.50
Sulphates	6.00
Chlorides	122.50
Carbonates.......	17.55
Water of crystallization..................... ...	7.60
	291 07
Less oxygen, equivalent of.......	27.57
Total solids...................	263.50

Warm Sulphur Spring of Rio Pajarito.

This noted sulphur spring is at Rio Pajarito, Taos county. The native people ascribe many virtues to the "sublime heal· ing powers" of its waters. Its temperature is 68° Fah.

The following analysis in parts of 100,000, made by Oscar Loew in 1875 is here given:

Sodium carbonate..........................	17.01
Calcium and magnesium carbonates	7.19
Sodium sulphate.....................................	14.60
Sodium chloride....... .... ,......................	9.11
Silica.........	trace
Potassa	trace
Lithia	trace
Total.....	47.91

A foot note by the chemist states ;that the water contains carbonic acid and hydrogen sulphide.

Socorro Thermal Springs.

At the foot of the eastern slope of the south end of Socorro mountain is located the water supply of the town of Socorro.

These waters come from a fractured zone or fault in the mountain at that point, oozing and bubbling out at several places. The chief flow is from the lower or big spring where a small reservoir has been constructed from which the water is conducted in a pipe to Socorro, about three miles to the northeast. O. R. Smith and the writer, in 1901, gauged the flow, and found the same to be approximately 700,000 gallons in twenty four hours, from the whole group. During the fall of 1898 an earthquake shock visibly increased the volume of water from its former flow. What effect the late seismic disturbances have had on the flow, if any at all, is not known. A series of measurements on the capacity of the spring would be interesting to the people of Socorro that they may be able to know how the present earth tremors are affecting their water supply.

The waters are alkaline in character and have a temperature of 93° Fah. From the nature of the volcanic rocks about the springs and deposits of silicious sinter, it would seem that these springs in former times, when they were young, possessed a much greater degree of activity than at present. Since silica is soluble in hot alkaline solutions the roughness, to the touch, of the adjacent rocks is chiefly due to the coating of the silicious sinter which has been brought up in solution by the hot waters.

No town in New Mexico, other than Socorro, has such advantages in a gravity supply of pure water. The analysis made by R. H. Case and H. T. Goodjohn of the New Mexico School of Mines in 1903, is as follows:

Constituents.	Grains per U. S. Gallon.
Calcium carbonate	3.1702
Magnesium carbonate	.6967
Potassium carbonate	2.0529
Potassium sulphate	4.7243
Sodium sulphate	3.0005
Sodium chloride	1.5658
Silica	1.6114
Iron and alumina	none
Organic matter	trace
Total solids	16.8218

Analysis of the water of Dog Lake spring of Estancia plain, as made by the New Mexico School of Mines:

Constituents.	Grains per U. S. Gallon.
Magnesium sulphate	556
Calcium sulphate	437
Potassium sulphate	65
Magnesium chloride	36
Alumina	1
Volatile matter	171

Sulphur spring at the head of Rocky Arroyo, Eddy county, as analyzed by E. M. Skeats of El Paso:

Constituents.	Parts per 100,000
Silica	0.75
Carbonate of lime	24.20
Alumina	0.25
Sulphate of lime	26.90
Magnesium sulphate	30 30
Sodium sulphate	1.77
Sodium chloride	1.76
Water of crystallization	4.54
Total	90.47

The gases are sulphuretted hydrogen and carbon dioxide. This is splendid suphur water, although the solids are less than usual.

Sulphur Springs near "bottomless lake" Roswell, Chaves county:

(E. M. Skeats, Analyst.)

Constituents.	Parts per 100,000.
Silica	1.00
Alumina	trace
Calcium carbonate	22.10
Calcium sulphate	350.30
Magnesium sulphate	116.40
Magnesium chloride	47.60
Potassium chloride	66.50
Sodium chloride	1915.00
Borax	trace
Total solids	2518.90

The gases are carbon dioxide and sulphuretted hydrogen; the latter gas is very heavy. This water has been used with apparrent good effects on rheumatic patients.

Sulphur Spring on Berrendo river, Chaves county:

(E. M. Skeats, Analyst.)

Constituents.	Parts per 100,000.
Silica	5.50
Alumina	1.00
Calcium carbonate	17.05
Calcium sulphate	72.40
Magnesium sulphate	35.25
Magnesium chloride	12.45
Sodium chloride	148.05
Water of crystallization	5.30
Total solids	297.00

Sulphuretted hydrogen is present to a moderate degree.

Spring on Peñasco, Chaves county:

(E. M. Skeats, Analyst.)

Constituents.	Parts per 100,000.
Silica	0.50
Alumina	0.50
Calcium carbonate	18.00
Calcium sulphate	21.95
Magnesium sulphate	21.20
Sodium chloride	3.85
Water of crystallization	trace
Total solids	66.00

Sulphuretted hydrogen—faint. This water is a good mild aparient.

Water Grove.

Jurassic (?) sandstone, San Juan Mesa, 60 miles north of Roswell.

(E. M. Skeats, Analyst.)

Constituents.	Parts per 100,000.
Silica	trace
Alumina	trace
Calcium carbonate	16.00
Magnesium carbonate	13.65
Potassium sulphate	54.45
Potassium carbonate	22.50
Sodium sulphate	120.00
Sodium chloride	22 60
Total solids.	249.20

This is a remarkable water in every sense and perhaps, it has no equal in New Mexico in gout and rheumatic affections.

Carlsbad Spring.

Above the town of Carlsbad about two miles are the famous mineral springs of the Pecos valley, which are widely known as the "Carlsbad Springs."

Formerly, the town of Carlsbad was known as Eddy, but was changed to its presnt name on account of the supposed similarity of its mineral waters to those of Carlsbad, Austria. The principal striking difference of these two waters is in the carbonate of lime and magnesium cholride. The famous Austrian spring is heavy in magnesium chloride with a very small amount of carbonate of lime; while the spring in Eddy county is destitute of the magnesium compound and contains a great deal of lime carbonate.

Carlsbad springs in Eddy county break forth on either side of the Rio Pecos, with a flow of approximately 5,500 gallons per minute.

The waters of these springs have become noted and recognized for their great medicinal values; especially, in cases of kidney trouble and dyspepsia.

The following is an analysis of the waters of the spring taken from Governor Otero's Report of 1903; the authority of the analysis is not given:

Constituents.	Grains per U. S. Gal.
Sulphate of soda (glauber's salts)	44.02
Chloride of sodium (common salt)	50.50
Sulphate of magnesia (epsom salts)	21.63
Sulphate of lime	17.40
Carbonate of lime	14.00
Silica	1.20
Iron and alumina	1.20
Carbonate of magnesia	2.05
Water of crystallization	3.25
Total solids	155.25

Quelites Mineral Spring.

This spring is located about two miles above the old Mexican village of Quelites, on the Antonio Sedillo grant and on the northeast side of the San Jose river, at the foot of some bluffs of massive Cretaceous sandstones.

The water is tepid and partakes of its high mineral qualities due to the subterranean source from the underlying saliferous "red beds;" the spring may, with propriety, be classed as a saline-soda spring. In addition to the great amount of solid matter, the water is highly charged with carbon dioxide. The contained iron at depth is a carbonate, but on reaching the surface and in coming in contact with the air, under released pressure, it is precipitated in the ferric state, in and about the spring.

The flow of the spring is now quite small (being in extreme old age) and comes from the top of a large mound of travertine, which it has build up by its own action in times past.

For stomach troubles and kindred ailments, this water seems to be especially beneficial.

The following is an analysis, made by Prof. Walter F.

Haines of Rush Medical College, Chicago, under date of February 21, 1885:

Constituents.	Grains per gal.
Sodium chloride	768.091
Sodium sulphate.	561.553
Sodium bicarbonate	13.605
Sodium bromide	.109
Sodium iodide	trace
Sodium borate	.654
Sodium phosphate	trace
Potassium chloride	16.528
Lithium.	.308
Barium sulphate	trace
Calcium carbonate	138.504
Magnesium	56.897
Iron	3.858
Manganese	trace
Alumina	.698
Silica	1.867
Total	1562.672

Gila River Hot Spring.*

This spring and others are situated on the upper Gila, and near Diamond Creek, in southwestern Socorro county. The water is very pure as shown by its low percentage of mineral or solid matter.

The following analysis was made by Wm. M. Courtis, of Detroit, Michigan, October 5, 1903:

Constituents.	Parts in 100,000
Sulphuric acid combined with lime and potash	3.420
Ferrous sulphate	.221
Calcium sulphate	2.829
Soda and potash carbonates	14.020
Silica	6.562
Total solids	27.052

*Temperature 100° Fah.

The mineral springs above given are among the more prominent ones in New Mexico, yet they constitute only a fractional part of the entire number that exist within the boarders of the territory.

The localities of a few of these springs are given briefly in the following.

Hot Springs.

Caballo springs (?) five miles from Fort McRae, Socorro county 136° Fah.

Hot springs about six miles north of Faywood station, Grant county; temperature 150° Fah.

Hot springs at copper mines of Rio San Francisco, western Socorro county; temperature 130° Fah.

On Diamond creek, near mouth, Socorro county; temperature 151° Fah.

Mineral Springs.

Three miles east of Gallup, McKinley county.

Eighteen miles east of Abiquiu, Rio Arriba county.

Five miles east of Ojo Azufre, McKinley county, fifteen miles west of Fort Wingate. Water alkaline.

East of Great ranch and three miles northeast of Las Vegas, San Miguel county. Alkaline and sulphuretted waters.

Ojo Azufre, twenty miles west of Fort Wingate, McKinley county, sulphuretted, etc.

Ojo Sarco, on Rio Grande, north of Santa Barbara, Taos county.

Soda Springs.

Three miles north of Ojo Caliente, Taos county.

On Salado creek, four or five miles south of San Ysidro, Sandoval county.

Four or five miles south of Carrizo valley, Socorro county.

Thirteen miles northeast of Isleta, Bernalillo county.

Stinking spring; ten miles northeast of Coolidge, Valencia county. Sulphuretted water.

About four miles southeast of Petaca, Rio Arriba county.

Sulphur Springs.

In Chusca valley, Rio Arriba county.

West of Mesa Lucera, Valencia county.

Five miles south of Taos, Taos county.

Between Peñasco and Mora, on the Rio Pueblo.

Warm Springs.

At the head of San Diego canyon, Rio Arriba county.

Ojo Caliente, Mimbres river, fifteen miles north of Mimbres postoffice, Grant county.

Ojo Caliente, twelve miles southwest of Zuñi, Valencia county.

Ojo Caliente, near Cherryville and Canyada Alamosa, Socorro county; temperature 85° Fah.

Ordinary Springs.

Patterson's springs, western Socorro county, at Patterson postoffice.

Horse springs, western Socorro county, northeast of Patterson about six miles.

Gallina spring, in Gallinas mountains, eastern Lincoln county.

Estancia, Antelope and Buffalo springs, in the Estancia plain, Torrance county.

Chico springs, on ex-senator Dorsey's ranch, twenty miles from Maxwell City, Colfax county.

Hermy springs, upper Pecos near Willis postoffice; summer resort.

Gallo spring, at San Rafael, south of Grant postoffice, Valencia county.

CHAPTER XXXVIII.

WELL (Artesian) AND RIVER WATERS.

It is not possible to enter into detail concerning the irrigation and potable (river and well) waters of the territory in the brief *resume*, which this volume is intended to cover. Such an exposition, as would do justice to the waters of the territory, relative to their importance as a factor in the present and future development of the commonwealth, would require a volume of several hundred pages. To water alone we must look to the rebuilding of the vanquished empires of the Montezumas in the southwest.

Geologically, the conditions for irrigation are favorable in a general sense; practically, can the problem be solved? We answer yes. As an illustration of what can be done in the arid west, we point with pride to the Pecos valley. The oldest settlers of that valley are now realizing, what seemed to them a quarter of a century ago, a *mirage* on the arid western plain.

New ideas are fast superseding the old; water is now being developed in large quantities; where formerly, none was suspected to exist. For example, the plains about Deming, the Jornado del Muerto, at Lordsburg, the artesian flows at Roswell, the deep well at Hachita and shallow wells of the Estancia plain, all demonstrate the existence, distribution, storage and circulation of the underground water of which our present knowledge is very meager.

Many of the potable well waters are as highly mineralized and efficacious, perhaps, as a number of the so-called mineral springs. For instance, the Fred Schmidt well in the eastern part of the Jornado plain is very strongly impregnated with epsom salts.

Again, the well of the El Paso and North Eastern Railway at Carrizozo, near the *mal pais*, is strongly sulphur, with free sulphuretted hydrogen. Numerous instances occur throughout the territory of these mineralized wells. As stated in the chapter on mineral springs, most all the waters in New Mexico are mineralized or heavily charged with materials in solution. This condition has on certain occasions been a great source of annoyance on the part of some of the railways in securing suitable water for the boilers of their engines. From the collection of analyses of various river and well waters, embodied in this chapter, an idea of their general character may be deduced.

Waters of the El Paso and Southwestern Railway Company.

(By kind permission of the Company.)

This railway follows along the southern boundary line of New Mexico. The source of the waters is from the north, mostly from the Mimbres district.

The region traversed by the water is mainly volcanic; that is, the hills are principally volcanic—the soils of the valleys are composed of the detritus of volcanic rocks and ash—mostly from comparatively recent eruptions.

The waters, as might be expected, are of an alkaline carbonate impregnation. Their approximate analyses as made by E. M. Skeats, in parts per 100,000, are as follows:

Name of well.	Pelea.	Mal-pais	Columbus.	Hachita.	Playas.
Silica, alumina, etc....	8.25	5.00	4.00	2.02	12.91
Lime carbonate........	12.00	5.26	10.00	3.00	4.00
Magnesium carbonate..	3.62	7.76	....	1.51	trace
Sodium carbonate.....	....	30.20	9.00	19.35	33.84
Sodium sulphate.	20.33	24.28	24.45	4.87	31.70
Magnesium sulphate...	3.45	....	....	....	....
Sodium chloride.	14.35	10.50	10.50	3.50	29 55
Total solids	62.00	83.00	90.00?	34.25	112.00
Depth of well in feet...	164	168	225	685*	168

* The Hachita well at 480 feet had the same total solids as at the extreme depth of 685 feet.

Pecos Valley Waters.

Through the kindness of Professor E. M. Skeats of El Paso, it was made possible to include in this volume the analyses of all the principal waters of the Pecos valley. To him, credit is due for all these analyses, which were made in his laboratory; and it is understood that he be given credit where the analyst's name is not specifically given for any particular analysis. All analyses are based on parts in 100,000, unless otherwise given:

Demmet's Lake, east of Roswell.

Silica.....	3.00
Alumina.........	5.00
Calcium carbonate..........	16.50
Calcium sulphate..........	162.40
Magnesium sulphate..........	76.80
Sodium sulphate..........	4.10
Sodium chloride..........	129.00
Calcium sulphide (?)..........	37.80
Water of crystallization...	11.50
Total solids..........	446.10

There are springs on the margin of this lake which give out sulphuretted hydrogen; the odor from most of them is very weak—the others strong. The water is used to a limited extent for irrigating and appears to give fair results.

Carlsbad Water Supply.

The supply is developed in Dark canyon by a series of wells and tunnels in the bed of the canyon. The wells are about sixty feet deep; and are four miles southwest of Carlsbad. The average analysis is as follows:

Silica..........	2.00
Alumina......	3.00
Calcium carbonate..........	21.00
Magnesium carbonate	4.18
Magnesium sulphate..........	8.31
Sodium sulphate..........	8.10
Sodium chloride.........	3.05
Water of crystallization..	3.86
Total solids..........	53.50

This water is from an underground stream which has its source from springs in the Guadalupe mountains. Fifteen miles up the canyon the water flows on the surface. The character of the water varies somewhat with the season and with rains.

BLUE SPRINGS.

Large head springs, in a lime conglomerate, feeding Black river:

Silica	1.50
Alumina	2.00
Calcium carbonate	15.20
Calcium sulphate	51.50
Magnesium sulphate	21.66
Sodium sulphate	5.30
Sodium chloride	3.21
Potassium nitrate	.41
Water of crystallization	19 22
Total solids at 100° C	120.00

RIO HONDO ARTESIAN (ROSWELL) BASIN.

Notice has just been received that the Secretary of the Interior has authorized the purchase of the Hondo reservoir site. This reservoir is to be constructed in Chaves county, on the Rio Hondo, twelve miles west of Roswell. This is a natural site and with small amount of embankment can be given a capacity of 40,000 acre-feet. Practically all the water that the Hondo will furnish in low-water years can be impounded at this point. It is possible to irrigate 15,000 acres.

	Rio Hondo Waters above Picacho.	At Roswell (clear)	Springs in Hondo (above Roswell.)
Silica	trace	......	......
Alumina	trace	......	......
Calcium carbonate	14.40)	20.00	9.50
Magnesium carbonate	2.52)		
Calcium sulphate	59.20)	81.20	29.10
Magnesium sulphate	25.65)		
Sodium chloride	9.66)	9.34	3.40
Sodium sulphate	4.22)		
Water of crystallization	3 85	10.46	16.00
Total solids	119.50	121.00	58.00

ROSWELL'S ARTESIAN WELLS.

Well No. 1. Depth 227 feet; temperature 70.5° Fah:

Silica and Alumina	3.55
Calcium carbonate	17.75
Calcium sulphate	16.10
Magnesium sulphate	15.56
Magnesium chloride	1.54
Sodium chloride	24.50
Total solids	79.00

Well No. 2.   Depth 192 feet; temperature 69° Fah:

Silica, alumina and water of crystallization......	8.65
Calcium carbonate...............................	14.00
Calcium and magnesium sulphates..........	34.00
Sodium chloride.....................	11.35
Total solids	68.00

Well No. 3.   Depth 225 feet; temperature 71° Fah:

Silica and aluminum...............	2.02
Calcium carbonate	19.20
Calcium and magnesium sulphates...............	37.80
Sodium chloride....	21.98
Total solids.......	81.00

Well No. 4.   Depth 150 feet; temperature 64.5° Fah:

	April 1896.	Jan. 1897.
Silica, alumina and water of crytalli-		
zation	4.80	6.95
Calcium carbonate.........	19.00	17.50
Calcium and magnesium sulphates....	48.35	39.50
Sodium chloride.....................	40.85	29.05
Total solids...................	113.00	93.00

Well No. 5.   Depth 331 feet:

Silica, alumina and water of crystallization.......	3.30
Calcium carbonate	17.50
Calcium and magnesium sulphate	57.20
Sodium chloride........................	38.00
Total solids	116.00

These wells from No. 1 to No. 5, inclusive, are in what is known as the Roswell basin, extending between north springs and south springs, with Roswell centrally located.

It is observed that the various springs compare favorably, both in analysis and temperature, with the different artesian wells.   Also, the level of the headwaters of the springs practically governs the level at which the artesian waters will flow.

Berrendo Artesian Basin.

This district lies immediately to the north of Roswell in the Berrendo basin; which is similar in all respects to that of the Hondo.

An analysis from an artesian well in this basin gives the following results:

Silica, alumina and water of crystallization	20.00
Calcium carbonate	19.00
Calcium and magnesium sulphates	67.00
Sodium chloride	138.00
Total solids	244.00

Berrendo spring water:

Silica	2.00
Alumina	1.00
Calcium carbonate	22.00
Calcium sulphate	39.30
Magnesium sulphate	28.05
Sodium sulphate	21.80
Sodium chloride	146.00
Water of crystallization	4.20
Total solids	264.35

This water contains some sulphuretted hydrogen. It is observed that the water is saline in character, but no difficulty has been experienced in irrigating with it and results are just as good as with the water of the Hondo basin.

FELEZ ARTESIAN (HAGERMAN) BASIN.

Well No. 6, at Hagerman; depth 800 feet:

Silica	1.50
Alumina	.50
Calcium carbonate	20.90
Calcium sulphate	19.20
Magnesium sulphate	23.25
Sodium sulphate	4.08
Sodium chloride	3.10
Total solids	72.53

The following are analyses of the head waters of the Rio Felez before sinking:

	Head Spring*	From Stream.
Silica	.50	1.20
Alumina		.40
Calcium carbonate	18.80	12.50
Calcium sulphate	11.11	14.58
Magnesium sulphate	9.23	12.25
Magnesium carbonate	1.85	2.03

Sodium carbonate........;.........	....	3.26
Sodium chloride..................	3.85	4.33
Water of crystallization..........	1.38	1.84
Total solids..............	46.72	52.39

*Head Spring temperature, 64° Fah.

MESCALERO ARTESIAN BASIN.

This basin lies on the east side of the Pecos river from Hagerman. The analysis of water from an artesian well 350 feet deep, gives the following:

Silica, alumina and water of crystallization.......	28.60
Calcium carbonate...............................	10.40
Calcium sulphate...........................	184.50
Magnesium sulphate.............................	66.75
Sodium chloride.......................	5.75
Total solids...............................	296.00

MOUND SPRING.

This spring is near the Rio Pecos, northeast of Hagerman:

Silica, alumina, etc	11.34
Calcium carbonate...	12.90
Calcium sulphate......................	185.00
Magnesium sulphate.............................	50.C0
Magnesium chloride.............................	1.26
Sodium chloride......................	8.80
Water of crystallization.	1/.70
Total solids......................	287.00

Mound spring is a very large one and is on top of a mound built up by the solids from the water and drifted sands; it is practically an artesian well. As seen from the analysis it is gypsum water, coming from the gypsiferous sands in the northern part of the Mescalero basin. The water raises good alfalfa.

WILLINGHAM'S ARTESIAN WELL.

This well, in the Mescalero basin, carries a very large percentage of gypsum. It lies to the northeast of Hagerman, in the vicinity of Mound spring, and is 450 feet deep. The following is the analysis:

Silica................................	4.00
Alumina.....	2.00
Calcium carbonate...............................	14.00

Fig. 49—CECILL'S WELL, Largest in Pecos Valley.

Calcium sulphate 200.00
Magnesium sulphate................................. 55.10
Sodium chloride.................................... 7.15
Water of crystallization........................... 8.25

    Total solids. 290.50

RIO PEÑASCO BASIN.

An artesian well 250 (?) feet deep gives the following:

Silica, alumina and water of crystallization. 17.03
Calcium carbonate................................. 16.15
Calcium and magnesium sulphates.................. 53.90
Sodium chloride.................................... 2.92

    Total solids................................. 90.00

Water from the Pecos river, by Gilbert's ranch:

Silica .... .50
Alumina... .50
Calcium carbonate................................. 18.00
Calcium sulphate.................................. 21.95
Magnesium sulphate. 21.20
Sodium chloride................................... 3.85

    Total solids................................. 66.00

This water contains traces of free sulphuretted hydrogen.

Head Spring, in conglomerate:

Silica 4.00
Alumina... 1.00
Calcium carbonate................................. 21.45
Calcium sulphate 34 40
Magnesium sulphate............................... 34.65
Sodium sulphate................................... 3.65
Sodium chloride................................... 2.65
Water of crystallization........................... 5.20

    Total solids................................. 107.00

These springs are west and above the McDonald well.

McDonald's artesian well: depth 144 feet; temperature 66.5° Fah. The analysis gives the following results:

Silica and alumina................................ 1.50
Calcium carbonate. 22.00
Calcium sulphate.................................. 19.05
Magnesium sulphate............................... 31.35
Sodium sulphate................................... 24.60
Sodium chloride. 2.34
Water of crystallization........................... 11.16

    Total solids 112.00

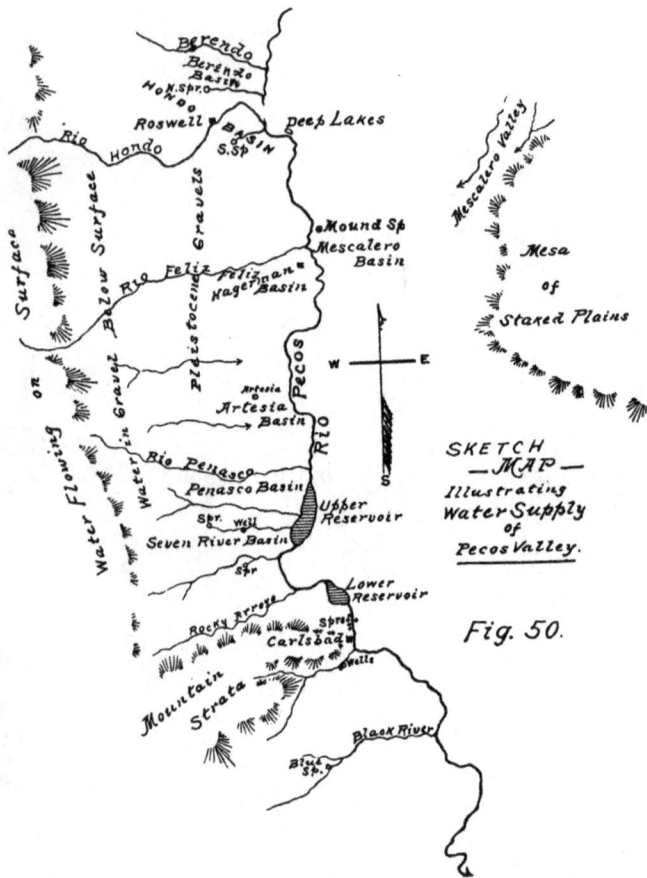

Berendo

Berendo
Basin

Ho N.Spr.o

Hondo

Roswell

Rio Hondo

BASIN

Deep Lakes

Surface

Rio Hondo

S.Sp.

Surface

Below Surface

Rio Feliz

Pleistocene Gravels

Water in Gravel

Mound Sp
Mescalero
Basin

Feliz
Hagerman
Basin

Pecos

Water Flowing

on

Artesia
Artesia
Basin

Rio Penasco

Penasco Basin

Rio Pecos

Mescalero Valley

Mesa
of
Staked Plains

W   E

S

Spr. Well

Seven River Basin

Spr.

Upper
Reservoir

SKETCH
—MAP—
Illustrating
Water Supply
of
Pecos Valley.

Lower
Reservoir

Rocky Arroyo

Spr.
Carlsbad

Fig. 50.

Mountain
Strata

Wells

Black River

Blue
Sp.

The Fanning spring is a large one and similar to the north and south head springs of Roswell; the temperature is 66.5° Fahrenheit:

Silica	3.00
Alumina..	2.00
Calcium carbonate.......	18.75
Calcium sulphate..	53.20
Magnesium sulphate............................	37.05
Sodium chloride..................................	3.51
Sodium sulphate.................................	9.09
Water of crystallization.....	5.40
Total solids......	132.00

Analysis of Rio Pecos Water.

(E. M. Skeats, Analyst.)

Constituents.	Above Santa Rosa.	Just above Roswell.	Upper reservoir.	Lower reservoir.	Above Carlsbad spring (Av.)	River at Pecos.	During first big flood of season	During second big flood of season.
SiO₂..........	1.75	} 6.00	5.00 }	3.00 }	} 3.00	6.00 }	3.00	3.00
Al₂O₃....... ..	.50			2.C0			2.00	6.00
CaCO₃........ ..	15.00	12.40	10.80	9.80	14 64	12.10	14.50	7.25
CaSO₄..........	144.30	181.00	} 192.00 }	96.83	35.U8	107.00	184.70	126.20
MgCO₃...	......	....		....	....	....	....	8.61
MgSO₄.....	33.00	69.30		95.65	29.00	67.98	57.30	23.40
Na₂SO₄........	....	2.50	20.80	50.82	40.00	30.20	10.85	1.27
NaCl..........	6.54	225.50	80.10	142.20	60.10	18/.50	111.20	49.26
Aq. of cryst...	21.91	23.80	25.30	7.70	20.18	58.22	.80	12.01
Total solids..	223.00	520.50	334.0u	408.00	202.00	464.00	384.35	237.00
In suspension..	clear	clear	clear	clear	clear	....	5,209.88	6.00

Below are given three analyses: The first is from a well (pumping station) in the Mesilla valley; the second is a sample of water from an acequia at Mesilla Park; the third is a sample of water from the Pecos river.

It is observed that the station well contains no suspended matter, whatever; while the Pecos river and especially the acequia carries a very large amount of silt. The analyses were made at the College of Agriculture and Mechanic Arts; the basis of analysis is parts per 100,000:

Constituents.	Station Well.	Acequia Water (Rio Grande.)	Pecos River.
Suspended matter............	......	831.40	179.70
Total solids....	104.00	44.11	312.59
Lime....	25.30	8.26	53.28
Magnesia......	5.65	1.36	17.08
Soda..........................	18.38	7.76	53.55
Potassa............................	2.13	.94	2.65
Iron and alumina....................	.....	1.85	.93
Silica	2.50	......	......
Sulphates	2.11	10.42	103.26
Chlorides.............	15.33	5.41	63.94
Carbonates	9.90	5.06	3.19
Cryst. Water of......	26.16	4.27	29.16
Totals................	107.46	45.33	327.04
Ogygen equivalent of Cl (deduct).....	3.46	1.22	14.45
Total solids (corrected).	104.00	44.11	312.59

Water Supply of Albuquerque.

The following analysis is furnished by the Water Supply Company of Albuquerque, which supplies the city with water:

(John Weinzirl, chemist, University of New Mexico.)

Ingredients.	Grains per U. S gallon.
Sodium chloride	1.03
Sodium sulphate............................	5.25
Sodium carbonate.......................	1.71
Calcium carbonate.....	3.71
Magnesium carbonate.....................	1.70
Silica..................................	4.92
Iron and alumina... ....·....	.25
Water of crystallization, etc...................	.18
Total solids......................	18.75

Reaction alkaline.

Hardness.	Degrees.
Temporary....................°.....................	2.58
Permanent............................	5.08
Total......	7.66

Analyzed July 13, 1093.

The total depth of the Water Supply Company's deepest well is 700 feet; but the main supply comes from a stratum of sand and gravel 185 feet below the surface. This well is in the northeast part of the city, in the Rio Grande valley; the flow is exceedingly strong. There are eight wells all told.

This is an excellent water for drinking and domestic purposes.

Mimbres Water (Deming).

The Mimbres River Water Company, which has recently been organized for supplying the city of El Paso with water from the under-flow of the Mimbres river, immediately south of Deming, has kindly furnished the following analysis from the Pittsburg Testing Laboratory:

Constituents.	Grains per U. S. gallon.
Free carbonic acid	1.54
Silica	4.32
Calcium carbonate	7.81
Magnesium carbonate	3.61
*Sodium oxide	8.28
Sodium sulphate	5.78
Sodium chloride	3.57
Sodium nitrate	.92

*This soda exists as sodium silicate and carbonate.

The above analysis is typical of the water supply of Deming.

Estancia Spring.

The following analysis of water from this spring, calculated under two different forms, was furnished by the Pennsylvania Development Company.

Constituents	Parts per 100,000	Grains per U. S. Gallon
Silica	2.50	1.45
Alumina	1.00	0.58
Magnesia	3.60	2.09
Soda	1.43	0.83
Sulphuric acid	2.79	1.62
Carbonic acid	9.69	5.63
Chlorine	1.64	.95
Lime	9.22	5.36
	31.87	18.51
Less oxygen equiv. of Cl	.37	.20
Total solids 120° C	31.50	18.31

The probable hypothetical combinations are as follows:

Silica..........	2.50	1.45
Alumina....................	1.00	0.58
Carbonate of lime.........	16.50	9.65
Carbonate of magnesium...	4.64	2.70
Sodium chloride...........	2.69	1.56
Water of crystalization...	4.17	2.37
Total.....	31.50	18.31

The water is good for railroad purposes, since it does not corrode iron to any appreciable extent. Most of the solids are precipitated on boiling and comes down as sludge, which does not stick to the boiler tubes.

Estancia Spring is at the new village of Estancia, a station on the line of the Santa Fe Central Railway, in Torrance county.

CHAPTER XXXIX.

TABLE OF ALTITUDES.

The following tabulated list of elevations, which embrace a number of the most prominent land marks and points in New Mexico, was taken mainly from House Documents Vol. 88 of the 55th Congress.

This volume is the third edition of the Dictionary of Altitudes, compiled by Mr. Henry Gannett, Geographer of the U. S. Geological Survey, under date of January 25, 1899. Some few points of elevation have been added since the time of that publication, which the Director or the Survey kindly furnished upon request.

The writer has interpolated some altitudes of his own which were made in the fall of 1893 with the spirit level, in making a railway survey from Maxwell City, a point on the Atchison, Topeka and Santa Fe Railway, westward to the Rio Grande, by way of Cimarron, to Moreno valley and Elizabethtown, over the Taos Pass, and through Taos. Two observations were made at Socorro; the spirit level was used for the Weather Bureau determination and trigonometric leveling for Socorro peak. Some other points in various localities were determined by the writer with the aneroid on different occasions.

Each point with its corresponding altitude and authority is here given:

STATION	AUTHORITY	ELEVATION
Abiquiu	Wheeler	5,930
Abiquiu Peak	Wheeler	11,240
Aden	S. P. R. R.	4,388
Afton	S. P. R. R.	4,204
Agua Azul	Wheeler	6,682
Agua Flegra	Wheeler	8,194
Agua Fria	Wheeler	6,486
Agua Fria Peak	Wheeler	10,965
Aguajes de los Guajolotes	Wheeler	6,202
Agua Negra	Wheeler	8,194
Alameda	A. T. &. S. F. R. R.	4,996

STATION	AUTHORITY	ELEVATION
Alamillo	A. T. & S. F. R. R.	4,651
Alamocita	Wheeler	4,256
Alamogordo	E. P. & N. E. R. R.	4,320
Alamosa	Wheeler	5,177
Alamo Station	Wheeler	6,593
Alannelo	Wheeler	4,693
Albert	Weather Bureau	4,700
Albuquerque	A. T. & S. F. R. R.	4,950
do	Weather Bureau	5,008
Alcalde	D. & R. G. R. R.	5,694
Aleman	Wheeler	4,594
Algodones	A. T. & S. F. R. R.	5,104
Alma	Weather Bureau	5,500
do	F. A. Jones	5,040
Alps	U. P., D. &. G. R. R.	6,508
Amargo	D. & R. G. R. R.	6,994
Animas Peak	Wheeler	6,106
Antelope Spring	Wheeler	6,221
Anthony	A. T. & S. F. R. R.	3,789
Anton Chico	F. A. Jones	5,370
Apache	Wheeler	7,324
Apache Cienega	Wheeler	5,291
Apache Tejo	Wheeler	5,478
Arch Spring	Wheeler	6,485
Arny	A. T. & S. F. R. R.	4,519
Atarque	Jones and Chadbourne	7,135
Atlantic & Pac. Junc.	A. T. & S. F. R. R.	4,891
Azotea	D. & R. G. R. R.	7,708
Aztec	Weather Bureau	5,590
Azul	A. T. & S. F. R R. (Bacon)	6,687
Bacon Spring	Wheeler	7,189
Baldy (Elizabeth Peak)	Wheeler	12,491
Baldy Peak	U. S. G. S.	12,623
Barranca	D. & R. G. R R.	6,934
Bayard, Fort	Wheeler	6,068
Bear Peak	Wheeler	8,081
Belen	A. T. & S. F. R. R.	4,801
Berenda Spring	Wheeler	7,494
Bernal	A. T. & S. F. R. R.	6,083
Bernal Hill	U. S. G. S.	7,020
Bernalillo	A. T. & S. F. R. R.	5,048
do	Weather Bureau	5,260
Bighorn	D. & R. G. R. R.	9,007
Black Mountain	Wheeler	8,909
Black Rock Tank	Wheeler	4,180
Blethen	U. P., D. & G. R. R.	8,162
Blossburg	A. T. & S. F. R. R.	6,857

STATION	AUTHORITY	ELEVATION
Bluewater	A. T. & S. F. R. R.	6,627
do	Weather Bureau	6,200
Blue Water Spring	Wheeler	6,778
Brookside	A. T. & S. F. R. R.	5,233
Buckman	Weather Bureau	8,500
Bueno Caballo	Wheeler	6,948
Burgwin Camp (old)	Wheeler	7,277
Burke	U. P., D. & G. R. R.	8,225
Burro Mountain (Big)	E. M. Chadbourne	7,175
Cabra Hill	U. S. G. S.	5,267
Caliente	D. & R. G. R. R.	7,309
Cambray	S. P. R. R.	4,221
Canyada Alamosa	Wheeler	6,540
Canyoncito	A. T. & S. F. R. R.	6,870
Canyon City	A. T. & S. F. R. R.	5,321
Canyon del Agua	Wheeler	5,916
Canyon del Chaco	Wheeler	5,839
Canyon Pajarito	Wheeler	5,099
Capitan Pass	Wheeler	7,398
Capitan Peak	Wheeler	10,023
Carrizillo Spring	Wheeler	4,457
Carrizo Peak	Wheeler	9,390
Carthage	A. T. & S. F. R. R.	5,026
Casa Colorado	Wheeler	4,679
Caseta	Wheeler	6,101
Catskill	U. P., D. & G. R. R.	7,872
Cerrillos	A. T. & S. F. R. R.	5,684
Cerro Blanco	Wheeler	14,269
Cerro Colorado	Wheeler	5,654
do	U. S. G. S.	6,800
Cerrro de Culebra	Wheeler	6,992
Cerro Pelon	U. S. G. S.	6,874
Cerro Tecolote	Wheeler	7,254
Chama	D. & R. G. R. R.	7,848
do	Weather Bureau	7,862
Chamisal	Wheeler	7,527
Chamita	D. & R. G. R. R.	5,626
Chase Camp	Wheeler	5,374
Chaves	A. T. & S. F. R. R.	6,987
Chico Spring	Wheeler	6,882
Chili	Wheeler	5,647
Ciboletta	Wheeler	6,411
Cibolo Hill	Wheeler	6,474
Cienega Armilla	Wheeler	7,179
Cimarron	Wheeler	6,385
Cieneguilla	Wheeler	6,011
Ciruela	Wheeler	6,744

STATION	AUTHORITY	ELEVATION
Clayton	U. P., D. & G. R. R.	5,054
do	Weather Bureau	5,178
Clemon	A. T. & S. F. R. R.	5,503
Cliff	F. A. Jones	4,522
Coleman	A. T. & S. F. R. R.	4,373
Colonas Ferry, Rio Grande	Wheeler	7,443
Colorado Mountain	Wheeler	5,654
Comanche	D. & R. G. R. R.	6,507
Comanche Canyon Pass	Wheeler	8,284
Cone Peak	Wheeler	12,690
Conrad, Fort	Medical Department, U. S. A.	4,576
Constancia	Wheeler	4,711
Continental Divide	A. T. & S. F. R. R.	7,243
Cook, Mount (Peak)	Wheeler	8,300
Cooke Spring	Mexican Boundary Survey	4,326
Coolidge	A. T. & S. F. R. R.	6,996
Cooney	F. A. Jones	5,840
Corazon Hill	U. S. G. S.	6,228
Corrales	Wheleer	5,091
Costilla Pass	Wheeler	10,188
Costilla Peak	Hayden	12,634
Cottonwood Springs	Wheeler	4,773
Covero	P. R. R. Reports	5,880
Cow Spring	Wheeler	5,001
Coyote Spring	Wheeler	7,202
Coyote Water Holes	Wheeler	6,775
Craig, Fort	Wheeler	4,448
Crawford	A. T. & S. F. R. R.	4,724
Cresco	D. & R. G. R. R.	9,178
Crocker	A. T. & S. F. R. R.	4,724
Cross Spring	Wheeler	6,265
Cubero	A. T. & S. F. R. R.	5,924
Cucamonga	Wheeler	5,954
Cuchilla Negra	Wheeler	4,568
Cuervo Hill	U. S. G. S.	5,309
Culebra	Wheeler	5,707
Cummings, Fort	Wheeler	4,778
Cutler	A. T. & S. F. R. R.	4,703
Datil Range, Western Peak	Wheeler	9,440
Defiance, Fort	Wheeler	7,042
Deming	Weather Bureau	4,315
do	S. P. R. R.	4,331
Desert	E. P. & N. E. R. R.	4,075
Des Moines	U. P., D. & G. R. R.	6,622
Detroit	A. T. & S. F. R. R.	4,008
Diablo Knoll	Wheeler	7,617
Dillon	A. T. & S. F. R R.	6,471

STATION	AUTHORITY	ELEVATION
Dog Canyon	E. P. & N. E. R. R.	4,025
Doña Ana	A. T. & S. F. R. R.	3,916
Dorsey	A. T. & S. F. R. R.	5,900
Dover	A. T. & S. F. R. R.	5,834
Dowlin Mill	Wheeler	6,435
Dripping Springs	Wheeler	5,623
Dulce	D. & R. G. R. R.	6,764
Eddy	Weather Bureau	3,122
El Cuervo Butte	U. S. G. S.	6,968
El Valle	E. P. & N. E. R. R.	5,380
Elizabeth Peak (Baldy)	Wheeler	12,491
Elizabethtown	Wheeler	8,465
Elk Spring	Wheeler	7,415
Elmoro	Wheeler	7,414
Elota	A. T. & S. F. R. R.	5,140
El Puerto de la Laguna	Wheeler	7,187
El Rito	A. T. & S. F. R. R.	5,657
Embudo	D. & R. G. R. R.	5,806
Emery Gap	U. P. D. & G. R. R.	6,462
Engle	A. T. & S. F. R. R.	4,762
do	Weather Bureau	4,750
Escobas Peak	Wheeler	8,278
Escondido	E. P. & N. E. R.R.	4,015
Española	D. & R. G. R. R.	5,575
do	Weather Bureau	5,590
Esteros	Wheeler	5,320
Eureka	Wheeler	4,746
Eureka Springs	Wheeler	4,239
Fairbell Hill	Wheeler	6,589
Florida	A. T. & S. F. R. R.	4,501
Florida Pass	P. R. R. Reports	4,600
Florida Peak	Wheeler	7,295
Folsom	U. P., D. & G. R. R.	6,399
Foraker's Ranch	U. S. G. S.	5,773
Fort Bayard	Weather Bureau	6,040
Fort Union	Weather Bureau	6,750
Fort Wingate	Weather Bureau	6,649
Fra Cristobal	Wheeler	6,646
Frisco	Jones & Chadbourne	5,900
Fulton	A. T. & S. F. R. R.	6,542
Gage	S. P. R. R.	4,480
Galisteo	Wheeler	6,117
do	Weather Bureau	6,074
Gallinas Peak	Wheeler	9,798
Gallinas Spring	Wheeler	7,673
Gallo Spring	Wheeler	7,587
Gallup	A. T. & S. F. R. R.	6,498

STATION	AUTHORITY	ELEVATION
Garcia Peak	Wheeler	9,920
Georgetown	Wheeler	6,455
Geyser Spring	Wheeler	5,400
Gila	Weather Bureau	4,040
Gila Hotel	U. S. G. S.	4,470
Glorieta	A. T. & S. F. R. R.	7,432
Gold Hill	F. A. Jones	5,770
Graham	U. S. G. S.	5,102
Grama	A. T. & S. F. R. R.	4,342
Grande	U. P. D. & G. R. R.	6,408
Gran Quivira	Wheeler	6,407
Grant	A. T. & S. F. R. R.	6,458
Greenville	U. P. D. & G. R. R.	5,952
Guadalupita	Wheeler	7,677
Gusano	A. T. & S. F. R. R.	7,243
Guy Fawkes	Wheeler	6,700
Hachita Peak	Wheeler	8,352
Hanover Peak	Wheeler	7,396
Hatch	A. T. & S. F. R. R.	4,050
Hedionda Lake	Wheeler	7,149
Hendrick Peak	Wheeler	7,574
High Peak	Wheeler	9.434
High Rolls	E. P. & N. E. R. R.	6,550
Hillsboro	Wheeler	5,224
do	Weather Bureau	5,100
Horse Springs	Wheeler	7,045
Hosta Butte	U. S. G. S.	8,837
Hot Springs	Wheeler	5,065
Hot Srings, Diamond Creek	Wheeler	5,545
Hudson	A. T. & S. F. R. R.	4,910
Hueco	E. P. & N. E. R. R.	4,033
Hurricane Rock	Wheeler	6,479
Indian Agency, near Blazier Hill	Wheeler	6,447
Isleta	A. T. & S. F. R. R.	4,898
Jarilla	E. P. & N. E. R. R.	4,170
Jemez Mountain	Wheeler	9,534
Jemez Peak	Wheeler	8,569
Jemez Pueblo	Wheeler	5,479
Jicarilla	U. S. G. S.	12,944
Johnson's Lake	E. M. Chadbourne	7,000
Juanita	D. & R. G. R. R.	6,326
Juan Lujan	Wheeler	6,011
Kelly	A. T. & S. F. R. R.	7,149
Kettle Spring	Wheeler	4,540
Kingman	A. T. & S. F. R. R.	6,821
Kiowa Spring	Wheeler	7,226
Labajada	Wheeler	5,515

STATION	AUTHORITY	ELEVATION
Labella	Weather Bureau	9,626
Labolsa	Wheeler	5,829
Lachusca	Wheeler	6,703
Ladrones Peak	Wheeler	9,214
La Glorieta	Wheeler	7,048
Laguna	A. T. & S. F. R. R.	5,786
Laguna del Ojo Hediondo	Wheeler	7,181
Laguna Gallinas	Wheeler	6,393
Laguno Los Griegos	Wheeler	6 956
Lajara Valley	Wheeler	6,988
La Joya	A. T. & S F. R. R.	4,702
Lake Peak	U. S. G. S.	12,380
La Lacha Spring	Wheeler	4,756
La Laguna de Sal	Wheeler	6,047
La Luz	E. P. & N. E. R. R.	4,836
La Monica	Wheeler	7,735
Lamy	A. T. & S. F. R. R.	6,475
Lanark	S. P. R. R.	4.162
Lansing	A. T. & S. F. R. R.	7,070
La Placita	Wheeler	5,129
Las Cruces	A. T. & S. F. R. R.	3,888
do	Weather Bureau	3,500
Las Lunas	A. T. & S. F. R. R.	4,848
Las Tapiecitas	Wheeler	8,810
La Tenaja	Wheeler	4,701
Las Tenajas	Wheeler	4,749
Las Truchas Mountain	Wheeler	13,150
Las Tusas	Wheeler	7,537
Las Vegas	A. T. & S. F. R. R.	6,398
Las Vegas Hot Springs	A. T. & S. F. R. R.	6,726
Laughlin Peak	Wheeler	8,950
Lava	A. T. & S. F. R. R.	4,720
La Veta	Wheeler	6,266
L. C. Ranch	F. A. Jones	4,500
Leidendorf Wells	Wheeler	6,401
Levy	A. T. & S. F. R. R.	6,255
Lisbon	S. P. R. R.	4,275
Llano	Wheeler	7,452
Llano Spring	Wheeler	5,348
Lobato	D. & R. G. R. R.	8,288
Lone Mountain	U. S. G. S.	7,310
do	Wheeler	5,986
Lordsburg	S. P. R. R.	4,242
do	Weather Bureau	4,245
Los Alamos	Wheeler	6,789
Los Brazos	Wheeler	7,321
Los Chaves	Wheeler	4,775

STATION	AUTHORITY	ELEVATION
Los Cornudos	Wheeler	4,362
Los Lunas	Weather Bureau	4,900
Los Machos	Wheeler	7,290
Los Ojos (Rio Chama)	Wheeler	7,273
Los Pinos	Wheeler	4,675
Los Quelites	Wheeler	5,134
Lower Peñasco	Weather Bureau	5,250
Luceros	Wheeler	7,941
Luera Springs	Wheeler	7,585
Lumber Spur	D. & R. G. R. R.	7,991
Luna	A. T. & S. F. R. R.	5,281
Lyndon	A. T. & S. F. R. R.	3,795
Lynn	A. T. & S. F. R. R.	7,529
McAlpine	U. P., D. & G. R. R.	7,937
McCarty	A. T. & S. F. R. R.	6,161
McEver Ranch	Wheeler	5,086
McRae, Fort	Wheeler	4,395
Macho or Mule Springs	Wheeler	5,261
Magdalena	A. T. & S. F. R. R.	6.557
Magdalena Pass	Wheeler	4,755
Magdalena Mount	Wheeler	10,798
Malpais Springs	Wheeler	4,106
Mangas Spring	Wheeler	4,799
Manuelito	A. T. & S. F. R. R.	6,252
Manzanares	A. T. & S. F. R. R.	6,586
Manzano	Wheeler	6,961
Manzano Peak	Wheeler	10,086
Marcy, Fort	Medical Department, U. S. A.	6,846
Martinez Mesa	Wheeler	6,820
Maxwell	A. T. & S. F. R. R.	6,078
Mesa Agua Segura	U. S. G. S.	6,044
Mesa Cherisco	U. S. G. S.	5,303
Mesa Huerfana	U. S. G. S.	6,455
Mesa Jacinto	U. S. G. S.	6,701
Mesa Pino	U. S. G. S.	5,502
Mescalero Agency	Wheeler	6,475
Mesquite	A. T. & S. F. R. R.	3,832
Mesteño	Wheeler	6,399
Metcalfe's Ranch	F. A. Jones	4,735
Mimbres Forks	Wheeler	7,563
Mimbres	Wheeler	4,920
Mimbres Mountain	Wheeler	10,061
Mimbres Settlement	Wheeler	5,007
Mogollon Creek (mouth)	U. S. G S.	4,650
Monero	D. & R. G. R. R.	7,263
do	Weather Bureau	7,356
Monica Springs	Wheeler	7,602

STATION	AUTHORITY	ELEVATION
Moolen Peak	King	7,339
Mora Canyon	Wheeler	6,528
Mosca Peak	Wheeler	9,723
Mound Spring	Wheeler	4,336
Mountain Key Mill	U. S. G. S.	6,569
Mount Dora	U. P. D. & G. R. R.	5,676
Mule Spring	Wheeler	5,282
Mule Springs	Wheeler	5,652
Nacimiento	Wheeler	7,300
Nacimiento Peak	Wheeler	10,045
Nambe Pueblo	Wheeler	6,045
Navajo	D. & R. G. R. R.	6,573
Neides Spring	Mexican Boundary Survey	4,310
New Placer	Wheeler	6,667
New York Mountain	Wheeler	10,594
Nigger Head Spring	Wheeler	4,861
No Agua	D. & R. G. R. R.	8,190
Nogal Peak	Wheeler	9,983
North Franklin Peak	Wheeler	7,070
Nutria	Wheeler	6,901
Nutria Spring	Wheeler	6,934
Nutrites Plaza	Wheeler	7,455
Nutt	A. T. & S. F. R. R.	4,706
Oak Spring	Wheeler	5,243
Oak Spring,Enciñoso	Wheeler	7,204
Ocate	A. T. & S. F. R. R.	5,939
do	Weather Bureau	7,500
Ocate Crater	Wheeler	8,903
Ojitos de las Cuevas	Wheeler	5,902
Ojo Amarillo	Wheeler	6,384
Ojo Caliente	Wheeler	6,292
Ojo Camaleon	Wheeler	6,401
Ojo Datil	Wheeler	7,419
Ojo de la Estancia	Wheeler	6,177
Ojo de Inez	Mexican Boundary Survey	5,293
Ojo de la Casa	Wheeler	6.243
Ojo de la Culebra	Wheeler	5,707
Ojo del Alto Peak	Wheeler	6,950
Ojo de la Parida	Wheeler	4,929
Ojo de la Quinca	Wheeler	5,973
Ojo de las Cañas	Wheeler	5,131
Ojo de la Tuñisa	Wheeler	5,673
Ojo del Cibolo	Wheeler	5,749
Ojo del Indio	Wheeler	9,280
Ojo de los Cazos	Wheeler	7,615
Ojo del Oso	Wheeler	7,786
Ojo de los Valles	Wheeler	6,979

STATION	AUTHORITY	ELEVATION
Ojo del Perro	Mexican Boundary Survey	4,692
Ojo de Nuestra Señora	Wheeler	6,606
Ojo de la Vaca	Mexican Boundary Survey	4,989
Ojo de la Vaca	Wheeler	6,964
Ojo del Gallo	Wheeler	7,943
Ojo del Milagro	Wheeler	5,173
Ojos Calientes	Wheeler	5,594
Onava	A. T. & S. F. R. R.	6,745
Organ Mountain	Wheeler	9,108
Organ Mountain Pass	P. R. R. Reports	5,467
Ortiz	A. T. & S. F. R. R.	5,836
Ortiz Mountains	U. S. G. S.	8,928
Oscura Water Holes	Wheeler	5,461
Osha Peak	Wheeler	10,223
Otero	A. T. & S. F. R. R.	6,394
Palmilla	D. & R. G. R. R.	8,241
Palomas	Wheeler	4,127
Paraje	Wheeler	4,319
Parida	Wheeler	4,627
Patero	Wheeler	6,123
Pecos	Emory	6,366
Pedernal Pass	Wheeler	7,181
Pedernal Peak	Wheeler	7,580
Pedernal Water Hole	Wheeler	7,140
Pelado Peak	Wheeler	11,260
Peña Blanca	Wheeler	5,170
Peñasco	Wheeler	7,452
Peralta	Wheeler	4,661
Pescado Spring	Wheeler	6,546
Picacho Crossing	Wheeler	3,784
Picacho de Sabinal	Wheeler	4,676
Picacho Peak	Wheeler	4,825
Picuris Pueblo	Wheeler	7,108
Pinos Altos	Wheeler	6,845
Pinos Altos	U. S. G. S.	7,041
Pinos Altos Peak	Wheeler	8,128
Pintado Pueblo	Wheeler	6,506
Pinto	E. P. & N. E. R. R.	5,953
Placer Mountain	Wheeler	8,827
Plaza del Alcalde	Wheeler	5,756
Plaza Mangoes	Wheeler	7,319
Pleasanton	U. S. G. S.	4,569
Point of Rocks	Wheeler	4,268
Pojoaque	Wheeler	5,750
Polvadero Peak	Wheeler	7,328
Poñil Pass	Wheeler	9,844
Pope	A. T. & S. F. R. R.	4,574

STATION	AUTHORITY	ELEVATION
Posos del Pino	Wheeler	6,055
Powell, Mount	U. S. G. S.	8,851
Pueblo Colorado	Wheeler	6,368
Pueblo Springs	Wheeler	6,363
Puertocito Spring	Wheeler	6,499
Puerto de Luna	Weather Bureau	5,350
Punta del Agua	Wheeler	6,599
Punta de la Mesa de San Jose	U. S. G. S.	7,667
Punta del Salitre	U. S. G. S.	6,719
Punta Pajarita	U S. G. S.	7,677
Pyramid	S. P. R. R.	4,298
Pyramid Hill	Wheeler	6,628
Quelites	Wheeler	5.193
Ralston	Wheeler	4.488
Randall	A. T. & S. F. R R.	3,964
Raton	A. T. & S. F. R. R.	6,637
do	Weather Bureau	6.400
Raton Tunnel	A. T. & S. F. R. R.	7.623
Rayado	Wheeler	6,946
Real Dolores	Wheeler	6,802
Recas	A. T. & S. F. R. R.	7,066
Red Bluff	P. V. R. R.	2,877
Red Rock	Jones & Chadbourne	4,150
Rincon	A. T. & S. F. R. R.	4,031
do	Weather Bureau	4,030
Rinconada	Wheeler	5,829
Rio Puerco	A. T. & S. F. R. R.	5,044
Rito Mangos	Wheeler	7,319
Rito Quemado	Wheeler	6,827
Roblado Peak	Wheeler	5,575
Rockspring	Wheeler	6,849
Rogers	S. P. R. R.	3,725
Romero	A. T. & S. F. R. R.	6,303
Rosario	A. T. & S. F. R. R.	5,417
Roswell	P. V. R. R.	3,565
do	Weather Bureau	3,570
Royce	U. P. D.. & G. R. R.	5,320
Sabinal	A. T. &. S. F. R. R.	4,758
Sabinal Agency	Wheeler	4,757
Salado	Wheeler	6,321
Salinas Peak	Wheeler	9,040
Salt Lake (crater)	F. A. Jones	6,500
San Antonio	A. T. & S. F. R. R.	4,534
San Antonio Peak	Hayden	10,833
San Antonio Valley	Wheeler	8,367
San Augustine	Wheeler	6.002
San Augustine Pass	Wheeler	5,654

STATION	AUTHORITY	ELEVATION
San Augustine Plain	Wheeler	6,780
Sandia Mountains	Wheeler	10,609
San Domingo	Wheeler	5,190
Sands	A. T. & S. F. R. R.	6,403
San Felipe	Wheeler	5,007
San Francisco Plaza, upper	Wheeler	5,688
do, " lower	Wheeler	5,639
do, " middle	Wheeler	5,638
do, Plaza, Rio Puerco	Wheeler	5,444
San Geronimo	Wheeler	6,724
San Ignacio	Wheeler	5,515
San Ildefonso	Wheeler	5,457
San Isidro	Wheeler	5,460
San Jose	A.. T. & S. F. R. R.	5,448
San Juan	Wheeler	5,601
San Lorenzo	Wheeler	6,107
San Lorenzo Spring	Wheeler	5,326
San Luis Rey	Wheeler	5,152
San Marcial	A. T. & S. F. R. R.	4,454
San Marcial ·	Weather Bureau	4,554
San Marcos Spring	Wheeler	6,056
San Mateo	Wheeler	7,323
San Miguel	A. T. & S. F. R. R.	6,036
San Nicolas Spring	Wheeler	4,218
San Pedro	Wheeler	4,488
San Pedro Mountains	U. S. G. S.	8,375
San Rafael	Wheeler	6,509
Santa Ana	Wheeler	5,346
Santa Clara Mountain	Wheeler	11,507
Santa Cruz	Wheeler	5,590
Santa Fe	A. T. & S. F. R. R.	6,954
do	Weather Bureau	7,013
Santa Fe Baldy Peak	Wheeler	12,661
Santo Niño del Rincon	Wheeler	7,418
Santa Rita Copper Mines	Wheeler	6,161
Santa Rita del Cobre	Mexican Boundary Survey	6,106
Sapello	Wheeler	6,876
Selden	A. T. & S. F. R. R.	3,954
Sellers	A. T. & S. F. R. R.	4,512
Separ	S. P. R. R.	4,500
Servilleta	D. & R. G. R. R.	7,713
Shattucks Ranch	Weather Bureau	6,000
Sherman Camp (old)	Wheeler	6,927
Shoemaker	A. T. & S. F. R. R.	6,271
Silver City	A. T. & S. F. R. R.	5,851
do	U. S. G. S.	5,933
Silver City Springs	Wheeler	7,638

STATION	AUTHORITY	ELEVATION
Slocum	Wheeler	4,519
Smith	U. P., D. & G. R. R.	7,591
Socorro	A. T. & S. F. R. R.	4,582
do	Weather Bureau (F. A. Jones)	4,600
Socorro Peak	Wheeler	7,281
do	F. A. Jones	7,258
Soledad	E. P. & N. E. R. R.	4,096
South Florida Peak	Wheeler	7,261
South Oscura Mountain	Wheeler	7,832
South Sandia Peak	Wheeler	8,567
Springer	A. T. & S. F. R. R.	5,783
do	Weather Bureau	5,857
Stanton, Fort, (flagstaff)	Wheeler	6,151
Steeple Bock	U. S. G. S.	5,133
Stinking Springs	Wheeler	6,249
Strauss	S. P. R. R.	4,080
Sublette	D. & R. G. R. R.	9,261
Sulzbacher	A. T. & S. F. R. R.	5,900
Sunday Peak	Wheeler	6,030
Sweetwater Spring	Wheeler	6,343
Tanques de Cañoncito	Wheeler	5,083
Tanques de las Animas	Wheeler	6,404
Taos	Wheeler	6,983
do	F. A. Jones	7,078
do	Weather Bureau	6,983
Taos Pass	Wheeler	9,282
	F. A. Jones	9,353
Taos Peak	Wheeler	13,145
Tapiacetas	Wheeler	8,870
Taylor, Mount	U. S. G. S.	11,389
Tecolote	A. T. & S. F. R. R.	5,861
Tetilla Peak	Wheeler	7,060
Thomas, Mount	Wheeler	11,275
Thompson Peak	U. S. G. S.	10,546
Thompson Spring	Wheeler	7,607
Thorn, Fort	Medical Department U. S. A.	4,500
Thunder Peak	Wheeler	9,122
Tierra Amarilla	Wheeler	7,466
Tijeras	Wheeler	6,214
Tipton	A. T. & S. F. R. R.	6,380
Toboggan Gulch	E. P. & N. E. R. R.	7,725
Toltec	D. & R. G. R. R.	9,450
do—Tunnel No. 2	D. & R. G. R. R.	9,622
Tomasceños Water Holes	Wheeler	5,502
Tome	Wheeler	4,879
Torreon Spring	Wheeler	5,980
Torres	U. S. G. S.	6,775

STATION	AUTHORITY	ELEVATION
Tres Cerros Spring	Wheeler	6,128
Tres Hermanas Peak	Wheeler	7,151
Tres Piedras	D. & R. G. R. R.	8,073
Triplets	Wheeler	4,347
Truchas	Wheeler	7,622
Truchas Peak	U. S. G. S.	13,275
Tularosa, Fort (old)	Wheeler	6,740
Tunica Mesa	Wheeler	5,510
Tunis	S. P. R. R.	4,419
Turquoise	E. P. & N. E. R. R.	4,103
Union, Fort	Wheeler	6,711
United States Mountain	Medical Department, U. S. A.	10,734
Upham	A. T. & S. F. R. R.	4,554
Upper Abo Pass	Wheeler	6,431
Ute Park	F. A. Jones	7,706
Ute Peak	Wheeler	10,151
Valencia	Wheeler	4,980
Valley Ranch	Weather Bureau	7,000
Valverde	A. T. & S. F. R. R.	4,486
Van Brummer	Wheeler	8,557
Vasquez	U. P., D. & G. R. R.	8,711
Vegas Village	Emory	6,418
Venado Spring	Wheeler	5,982
Vermejo	Wheeler	7,823
Vincent, Camp	Wheeler	6,188
Volcano	D. & R. G. R. R.	8,872
Wagon Mound	A. T. & S. F. R. R.	6,193
Waldo	A. T. & S. F. R. R.	5,621
Wallace	A. T. & S. F. R. R.	5,263
Warm Spring	Wheeler	5,008
Water Canyon	A. T. & S. F. R. R.	6,008
Watrous	A. T. & S. F. R. R.	6,413
Watson	A. T. & S. F. R. R.	4,510
Webster, Fort	Medical Department, U. S. A.	6,350
West Gallinas Mountains	Wheeler	8,464
West Jicarilla Cone	Wheeler	7,727
White Sands	Wheeler	3,888
White Oaks	Weather Bureau	6,470
Whiteoak Spring	Wheeler	6,618
Whitewater	A. T. & S. F. R. R.	5,151
Whitney	A. T. & S. F. R. R.	4,367
Wilder	U. P., D. & G. R. R.	7,848
Willow Spring	Wheeler	6,677
Wilna	S. P. R. R.	4,554
Wingate	A. T. & S. F. R. R.	6,736
Wingate, Fort	U. S. G. S.	6,997
Winsor's Ranch	Weather Bureau	8,000

STATION	AUTHORITY	ELEVATION
Winter Spring	Wheeler	7,182
Wooten	E. P. & N. E. R. R.	7.071
Yucca, Camp	Wheeler	4,374
Yule Spring	Wheeler	5,925
Zuñi	S. P. R. R.	4,184

CHAPTER XL.

CENSUS OF NEW MEXICAN MINERALS.

Mineral	Common Name	Locality
Alabandite	Manganese sulphide	San Pedro copper mines
Alabaster	Oriental onyx-marble	Jones mining district
Actinolite	Mg.-lime-iron amphibole	Jones iron mines, Sierra Oscura
Agate	Water stones	San Mateo and Sangre de Cristo Mts.
Albite	Soda feldspar	Various mountain ranges
Almandite	Iron-aluminum garnet	San Pedro copper mines
Allophane	Incrusted hyd. Al. sil.	Fierro and Hanover
Alunogen	Alum	Upper Gila and elsewhere
Amethyst	Amethystine quartz	Black range, Great Republic mine
Andradite	Iron-lime garnet	Organ and Cieneguilla districts
Anglesite	Sulphate of lead	Victorio mining district
Anhydrite	Dehydrated gypsum	Metamorphosed by lava flows
Anorthite	Lime feldspar	Various mountain ranges
Anthracite	Hard or stone coal	Madrid mines, Cerrillos district
Anthraconite	Stinkstone	Tejon and Quelites grants
Apatite	Phosphate of lime	Lake Valley and Hillsboro regions
Aragonite	Lime spar	Kingston camp and Graphic mine
Argentite	Silver glance	Kingston and Bromide district
Arsenopyrite	Mispickel	Virginia mining district
Asbestus	Mineral fiber	Mimbres mining district
Asphaltum	Asphalt, mineral pitch	Perea grant, Leonard Wood county
Aurichalcite	Copper-zinc carbonate	Magdalena mining district
Automolite	Spinel (form of)	Santa Fe mountains
Azurite	Blue carb. of copper	Santa Rita copper mines
Barite	Baryta, heavy-spar	Mimbres mining district
Baryto-calcite	Carb. barium and lime	Sierra Oscura
Baryto-celestite	Sulphate of Ba. & Sr.	Sierra Oscura
Biotite	Black mica	Ojo Caliente, Rio Arriba county
Bismuthinite	Bismuth trisulphide	Wilcox mining district (?)
Bloedite	Hyd.-Mg. & Na. Sulphate	Estancia salt lake
Bornite	Peacock copper	Black range and Cooney districts
Bournonite	Wheel ore	Los Cerrillos and Central districts
Brochantite	Basic copper sulphate	Torpedo mine, Organ district
Bromyrite	Silver bromide	Bromide (Tierra Blanca) district
Brookite	Rutile	Copper Mountain district
Bucklandite	Iron epidote	San Pedro copper mines
Calamine	Silicate of zinc	Thunderbolt mine and Magdalena Mts.
Calaverite	Telluride of gold	La Belle and Red River districts
Calcio-celestite	Sulphate of Ca. & Sr.	Sierra Oscura
Calcite	Calc-spar, limestone	Kelly and Graphic mines
Caliche	Lime deposit (Mexican)	Common to New Mexican plains
Carnelian	Red chalcedony	Found in various gravel beds
Carnotite	Uranium oxide (yellow)	Peralta canyon, Cochiti district
Catlinite	Indian pipe clay (red)	Sangre de Cristo range (?)
Celestite	Strontium sulphate	Sierra Oscura and San Andreas Mts.
Cerargyrite	Horn silver	Lake Valley and Kingston mines
Cerium	One of the rare metals	Gravels of Rio Chama (?)
Cerussite	Lead carbonate	Cooks Peak and Magdalena districts
Cervantite	Oxide of antimony	Central district, Grant county
Chalcedony	Waxy quartz (in luster)	Widely distributed over New Mexico
Chalcocite	Copper glance	Santa Rita and Cooney mines
Chalcopyrite	Pyrites of copper	Cooney mine and elsewhere
Chalcotrichite	Rare copper oxide	Santa Rita mines
Chert	Flint	Common in gravel beds
Chrysocolla	Silicate of copper	Torpedo mine, Organ district
Chrysolite	Olivine (peridot)	Zuñi Indian reservation
Coal	Coal	Occurs in Cretaceous rock system
Copper	Native	Santa Rita Mines
Corundum	Sapphire	Found in Santa Fe gravels
Covellite	Blue copper sulphide	Jarilla mountains
Cuprite	Red oxide of copper	Santa Rita mines
Cyanite	Silicate of aluminum	Bromide district, Rio Arriba county
Cymatolite	Albite-muscovite	Ojo Caliente, Rio Arriba county
Descloizite	Vanadium-lead-zinc oxide	Lake Valley

Mineral	Common Name	Locality
Diamond	Diamond (pure carbon)	Said to have been found (?)
Diaptose	Emerald copper (silicate)	Jarilla mountains
Dolomite	Magnesium limestone	Various mountain ranges
Domeykite	Copper arsenide	Pinos Altos and Central districts
Embolite	Chloro-bromide of silver	Lake Valley mine
Emerald	Beryl (gem)	Santa Fe gravel beds
Emeraudine	Same as diaptose	Jarilla mountains
Enargite	Copper arsenide	Pinos Altos and Central districts
Endlichite	Vanadiferous mimetite	Lake Valley and Hillsboro
Epidote	Pistacite (yel-green var)	San Pedro copper mines
Epsomite	Epsom salt	Estancia lakes
Erubescite	A type of bornite	Black range, Cooney mine
Euclase	Gem	Reported found (?)
Fiorite	Silicious sinter	Faywood and Socorro springs
Flint	Same as chert	Various localities
Fluorite	Flour-spar	Gila River and Sierra Oscura
Franklinite	Iron-zinc-mang. oxide	Central mining district
Gahnite	Variety of spinel	Cerrillos mining district
Galenite	Galena	Cooks Peak and Magdalena districts
Glauberite	Sodium-calcium sulphate	Estancia lakes
Gold	Gold, placer and lode	Elizabethtown and Hillsboro
Graphite	Black lead (plumbago)	In Raton and Sandia mountains
Grossularite	Lime-aluminum garnet	San Pedro and Organ mountains
Guano	Bat deposit (phosphate)	Extinct crater, near Engle
Gypsum	Lime sulphate (yeso, Mex.)	Ancho and plains of "white sands"
Halite	Rock (common) salt	Estancia and crater salt lakes
Hausmanite	Oxide of maganese	San Lorenzo mining district
Hematite	Specular (red oxide) iron	Jones iron mines and elsewhere
Hubernite	Tungsten (wolfram) ore	Victorio mining district
Hyalite	Opal family	Central and Cochiti districts
Hydro-zincite	Zinc bloom	Magdalena mining district
Ilmenite	Black iron ore	Fierro and Hanover
Ilvaite	Diprismatic iron ore	Fierro and Hanover
Idocrase	Vesuvianite	San Pedro and Organ Mts.
Iodyrite	Silver iodide	Lake Valley
Iridium	Metallic	Hillsboro
Jadeite	Jade, ornamental stone	Jicarilla mountains
Jasper	Brown iron stone (flint)	Canyoncito district
Jet	Jet (for jewelry)	Vicinity of Santa Rosa
Kaolinite	China clay (impure)	Socorro mountain
Labradorite	Lime-soda feldspar	Various mountain localities
Lead	Found native (rare)	Magdalena mountains
Lepidolite	Pink lithium rock	Cieneguilla mining district
Lieverite	Same as ilvaite	Hanover and Fierro
Lignite	Brown coal	Gallup and elsewhere
Limonite	Brown iron ore	North end of the Sandias
Magnetite	Lode stone	Fierro
Malachite	Carbonate of copper	Santa Rita and Las Vegas
Manganite	Manganese sesquioxide	Lake Valley
Manganosite	Black oxide of manganese	San Lorenzo mining district
Marble	Lime carbonate	Near Alamogordo
Marcasite	White iron pyrites	Manzano mountains
Marionite	Hydro-zincite (calamine)	Graphic mine
Massicot	Oxide of lead	Chloride flat
Melaconite	Black oxide of copper	Black range and Santa Rita
Melanotekite	Lead-iron silicate	Hillsboro, Las Animas district
Miargyrite	Antimony-silver sulphide	Kingston and Palomas camps
Mica	Isinglass	Petaca and Nambe
Microcline	Orthoclase, potash felds.	Various mountain ranges
Millardite	Hyd. maganese sulphate	Lake Valley mine
Mimetite	Vanad. arsenic chloride	Socorro mountain
Minium	Red oxide of lead	Hachita (Eureka mining district)
Mirabilite	Glauber salt	Estancia lakes
Molybdenite	Molybdenum trioxide	San Miguel county
Monazite	Phos. of the cerium earths	In Chama river sands (?)
Moonstone	Feldspar silicate	San Mateo mountains
Muscovite	White mica (isinglass)	Petaca, Nambe and Talco
Nickel	Metal	Upper Pecos region
Novaculite	Whetstone	Sangre de Cristo mountains
Obsidian	Volcanic glass	Santa Fe mountains
Ocher	Paint rock	Sandia mountains and San Pedro
Octahedrite	Titanium oxide (rutile)	Central mining district
Odontolite	Bone turquoise	Nacimiento mountains
Oligoclase	Soda-lime feldspar	Various mountain regions

Mineral	Common Name	Locality
Opal	Fire opal (gem)	Cochiti and Central districts
Orthoclase	Potash feldspar	Various mountain regions
Pectolite	Radiated fibrous silicate	Cieneguilla district
Peridot	Gem (crysolite, olivine)	Zuñi Indian reservation
Petalite	Lithium ore	Cieneguilla, Copper Mountain dist.
Petroleum	Kerosene (coal oil)	Vicinity of Gallup and San Juan Co.
Petzite	Gold and siver telluride	Lookout mine and La Belle dist.
Phlogopite	Brown mica (sparingly)	Nambe
Pistacite	Yellowish-green epidote	Red river
Platinum	Metallic	Tampa mine, Bromide No. 2 district
Plattnerite	Lead dioxide	Cooks Peak and Central districts
Polybasite	Silver-antimo. sulphide	Telegraph mining district
Proustite	Ruby silver	Kingston and Bullard's Peak
Przibramite	Cadmiferous sphalerite	Carpenter mining district
Psilomelane	Manganese dioxide	Near Rincon. Caballo mountains
Pumice	Filter stone	Near Grant P O. and Socorro
Pyrargyrite	Ruby silver	Bullard's Peak and Kingston
Pyrites	Cube iron	Various mining districts
Pyrolusite	Black iron (mang. dioxide)	San Lorenzo mining district
Pyromorphite	Lead phosphate	Macho district
Pyrostilpite	Ruby silver	Kingston and Bullard's Peak
Pyrrhotite	Iron sulphide	Fierro and Hanover
Quartz	Oxide of silicon	In various mines
Rhodocrosite	Manganese carbonate	Graphic and Kelly mines
Rhodonite	Manganese spar (fowlerite)	San Lorenzo mining district
Ricolite	Rich stone (ornamental)	Gila river, near Red Rock
Ruby	Gem	Reported found at Zuñi (?)
Rutile	Titanium dioxide	Central district
Satin spar	Fibrous gypsum	East of Strawberry Peak, Socorro
Scheelite	Calcium tungstate	Victorio mining district
Selenite	Transparent gypsum	Pittsburg district, Caballo Mts.
Siderite	Spathic iron ore	Granite Gap mine
Silver	Native	Silver Cell mine
Smithsonite	Zinc carbonate	Magdalena mining district
Sphalerite	Zinc blende (jack)	Cerrillos and Carpenter districts
Spodume	Lithium rock (petalite)	Cieneguilla district
Staurolite	Cruciform	Cieneguilla and Copper mountain
Steatite	Soapstone	Various localities
Stephanite	Brittle silver	Bromide mine, Rio Arriba county
Sternbergerite	Silver-iron sulphide	Cooney mining district
Stibnite	Sulphide of antimony	Cerrillos mining district
Stilbite	A hydrous silicate	Baldy mountain, Colfax county
Sulphur	Native	Otero Sulphur springs
Sylvanite	Telluride of silver-gold	Trujillo Creek, Sierra coun'y
Talc	Soapstone (var)	Numerous localities
Tennantite	Copper-arsenic sulphide	Pinos Altos and Central dists.
Tenorite	Black oxide of copper	Santa Rita mines
Tetrahedrite	Copper-arsenic sulphide	Pinos Altos and Central dists.
Titanite	Same as rutile	Central mining district
Torbernite	Uranium ore (rare)	Jerome mine, San Lorenzo dist.
Tourmaline	Complex silicate	Bromide District No. 2
Travertine	Geyser formation	Salt Lake crater, Socorro county
Tremolite	Amphibole, hornblende	Various mountain ranges
Tripolite	Infusorial earth	Reported near Socorro (?)
Troostite	Silicate of zinc	Magdalena mining district
Tufa	Volcanic mud	In eruptive regions
Turquoise	Used as a gem	Cerrillos and Burro districts
Uranophane	Rare uranium ore	Jerome copper mine, Socorro coun'y
Vanadinite	Vanadate of lead	Georgetown and Cerrillos
Vesuvianite	Idocrase	San Pedro Copper mine
Wad	Bog manganese ore	Central mining district
Willemite	Zinc silicate (troostite)	Merritt mine. Socorro mountain
Witherite	Barium carbonate	Sierra Oscura and San Andreas
Wolframite	Iron-mang. tungsten	Victorio mining district
Wollastonite	Tabular spar	San Pedro and Organ mountains
Wulfenite	Lead molybdate	Stephenson-Bennett mine
Xanthoconite	A sulpharsenite	Cerrillos mining district
Zaratite	Emerald nickel	San Miguel county
Zoisite	Lime epidote	San Pedro and Organ mountains
Zincite	Zinc oxide	Magdalena mountains

The foregoing list embraces all of what might with propriety be termed the *metallic* minerals of New Mexico; so far as known at the present time. This catalogue, also, includes a number of the more common mineral species of the silicates, feldspars and other distinctive types of mineral aggregations. Concerning these latter classes, no attempt was made to make the list complete; since such types would refer more to lithology than to mineralogy.

It is believed that the metallic list, proper, will be greatly augmented, by the discovery of most all of the rarer metals, when a careful and systematic study is given to the ores of the various mining districts.

So far as known, this is the first attempt ever made to catalogue the known *metallic* minerals of the territory.

Statistics.

Of late years, New Mexico has been on the decline in her metallic outputs, with the exception of copper, iron and zinc.

It might be said that copper is holding its own; iron is decidedly on the increase; and zinc has taken a phenomenally high spurt over anything in its previous history. In fact, the year 1904 will give a greater production of zinc than the aggregate of all previous years. The increase is due to the large bodies of zinc carbonate recently developed in the Magdalena district.

The resources usually regarded as non-metallic are all steadily on the increase; as also the manufacture of brick and lime. The cement plasters and allied products, manufactured at Ancho, and the new plant just completed at Roswell, give an impetus in the increased use of the raw material required in such manufacture.

Salt and sulphur are also forging to the front.

Coal stands at the head of the list in New Mexico's mineral resources, and the output is steadily advancing at a very healthy rate of increase.

As near as can be estimated, the production of gold in New Mexico from the earliest time down to January 1, 1904, is approximately $26,700,000; that of silver will approximate $29,000,000.

According to the Director of the Mint the production of

precious metals in 1902, was gold 25,693 ounces, valued at
$531,100; silver 457,200 fine ounces, valued at $591,127.

The statistics of precious metal production for 1902, as given
by the United States Geological Survey and also by the
Census Bureau make the amount considerably less than the
Mint report, which latter report is based on estimates.

The writer had charge of the Mining Census of New Mexico during the year 1902, in connection with the U. S. Geological Survey as Field Assistant, and herewith gives the data,
as collected by personal visits into the several camps and districts at that time:

Kind of Mine	Value of Production
Coal	$1,500,230
Gold....... ...,...............................	486,545
Silver ....:	190,623
Copper ....:............	271,270
Precious stones......	51,600
Quartzites	12,291
All other minerals	173,914
Total value.........	$2,686,473

It is seen that the production of coal is greater than all of
the other mineral products combined.

Producing Mines	Number
Coal........	30
Gold and silver (including placers)....................	91
Copper	17
Precious stones............................	8
Quarries	7
All others	8
Total producing mines........................	161

Number of mine operators..... 207
Number of salaried officials 175

Kind of Mine	No. wage earners	Earnings
Coal	1439	$1,027,460
Gold and silver..	519	409,779
Copper	164	128,483
Precious stones	36	22,087
Quarries	8	6,515
All other mines....	109	52,509
Total	2275	$1,646,833

APPENDIX.

Synopsis of the Mining Laws of the Territory of New Mexico Governing the Location and Relocation of Mining Claims.

(By M. E. Hickey.)

DIMENSION OF LODE CLAIM.

A lode claim may be 600 feet in width by 1,500 feet in length.

By Section 2324, Revised Statutes of the United States, the miners of each mining district may make regulations not in conflict with the laws of the United States, or with the laws of the State or Territory in which the district is situated, governing the location, manner of recording and the amount of work necessary to hold possession of a mining claim.

The local regulations of some of the mining districts in New Mexico make the lode claim 300 feet in width and 1,500 feet in length.

DIMENSION OF PLACER CLAIM.

A placer claim may contain twenty acres.

The following references are to the Compiled Laws of New Mexico, except where otherwise noted.

Section 2286. It is necessary in locating a mining claim, to distinctly mark the boundaries of such claim, and to post in some conspicuous place on such location a notice in writing giving such a description as will identify the claim, and the names of the locator or locators, and his or their intention of locating said claim.

A copy of said notice must be recorded in the office of the recorder of the county in which said claim is located, within three months of the posting of such notice. No other record of said notice is necessary.

Section 2299 (as amended by the Session Laws of 1899, Section 1, Page 111). Within 120 days of the location of any mining claim in this territory, the surface boundaries of such claim shall be marked by four substantial posts or monuments, one at each corner of said claim.

Section 2300. The re-location of any mining claim which is subject to re-location, shall be made in the same way as an original location is required by law to be made.

Section 2301. If the original location is defective in any way or the requirement of law has not been complied with before filing, or if the owner of any mining claim shall be desirous of changing his surface boundaries, or of taking in any part of an overlapping claim which has been abandoned, such owner may file an amended or additional notice of location, provided such notice does not interfere with the rights of others.

Section 2302. Any person who shall take down, remove, alter or destroy any stake, post, monument or notice of location upon any mining claim without the consent of the owner or owners thereof, shall be deemed guilty of a misdemeanor, and on conviction, shall be punished by a fine not exceeding one hundred dollars or by imprisonment in the county jail not exceeding six months, or by both such fine and imprisonment.

Section 2311. Any person or persons, or the manager, officer, agent or employe of any person, firm, corporation or association, who shall in any manner alter, deface or change the location notice of any mining claim in this Territory, located under the laws of the United States or of the laws of this Territory, or any location regulation in force in the district wherein such claim is situated, thereby in any manner affecting the rights of any person, firm or corporation, to such claim or location, or the land covered thereby, shall be deemed guilty of a misdemeanor, and upon conviction thereof before any court of competent jurisdiction, shall be fined in a sum not less than one hundred dollars, nor more than five hundred dollars, or imprisonment in the county jail for not less than sixty days, nor more than one year, or by both such fine and imprisonment, in the discretion of the court trying the case. Nothing herein contained shall affect the rights of such locator or locators, to correct errors in such notices as provided in section two thousand three hundred and one, and the laws of the United States: Provided, such change shall not affect or change the date of such location notice, or affect the right of any other person.

Section 2298. The locator or locators of any mining claim shall within ninety days from the date of taking possession of the same, sink a discovery shaft, exposing mineral in place, or shall drive a tunnel, adit, or open cut upon such claim to a depth of at least ten feet below the surface.

Section 2300 (latter part of the section)....The re-locator of a mining claim shall sink a discovery shaft, exposing mineral in place or shall drive a tunnel, adit, or open cut upon such re-located claim to a depth of ten feet below the surface, or shall sink the original shaft ten feet deeper, or drive the original tunnel, adit, or open cut upon such claim ten feet further.

Section 2315. The owners of unpatented mining claims in New Mexico shall within sixty days from and after doing the assessment work which is required by law to be done upon said claim, cause to be filed with the recorder of the county in which such mining claim is situated, an affidavit setting forth the fact that all work required by law has been done.

The annual assessment work must be done each year not later than the 31st day of December (Rev. Stats. U. S.).

On each claim, until patent is issued therefor, not less than one hundred dollars worth of labor shall be performed or improvements made during each year.... When five hundred dollars worth of labor has been performed, or improvements made, on any mining claim, the owner can get a patent from the United States for said claim upon compliance with the laws governing the issuance of patents, and upon payment to the United States of five dollars an acre for the land contained in said claim. (Rev. Stats. U. S.).

BIOGRAPHICAL.

MIGUEL A. OTERO, Governor

MIGUEL ANTONIO OTERO.

Governor M. A. Otero of New Mexico is a descendant of an old Spanish family. His father who bore the same Christian name, represented New Mexico in Congress. Governor Otero was born at St. Louis on October 17, 1859. He received his education at St. Louis University and the University of Notre Dame. On December 18, 1888, he married Caroline V. Emmett, daughter of ex-Chief Justice Lafayette Emmett, of Minnesota. Mr. Otero was cashier of the San Miguel National Bank at Las Vegas from 1880 to 1885. During that time he served as treasurer of the city. He was elected probate clerk of the County of San Miguel and served until 1890 after which he served three years as clerk of the United States and territorial district courts of the Fourth Judicial District of New Mexico. Governor Otero was a delegate to the Republican national conventions of 1892, 1900 and 1904, being chairman of the New Mexico delegation the latter two times. On June 7, 1897, he was appointed Governor of New Mexico by President McKinley. He was reappointed on June 15, 1901, by President McKinley and reappointed by President Roosevelt on December 18, 1901. He approved the bill creating the St. Louis Board of Louisiana Purchase Exposition Managers and making an appropriation of $30,000 for a New Mexico exhibit, passed by the 35th legislative assembly, March, 1903.

CHAS. A. SPIESS, President.

CHARLES A. SPIESS.

Charles A. Spiess, president of the New Mexico Board of Managers of the Louisiana Purchase Exposition, was born in Johnson county, Missouri, on March 19, 1867. He attended the public schools of his native state and graduated with honors from the Missouri Normal College. In 1888 he came to Las Vegas, San Miguel county, being employed as a clerk. Soon afterwards he removed to Mora county and served one year as deputy probate clerk and assessor. Removing to Santa Fe, he completed his law studies begun several years before, was admitted to the bar in 1891 and engaged in the practice of law. In 1893, 1894 and 1895, he served as a member of the board of education of the City of Santa Fe. In 1895 he was nominated to the legislative council by the Republicans of Santa Fe county but was defeated by 13 votes out of a total number of 3,300 cast. In 1896 he was again nominated and was elected.

Governor Otero appointed him district attorney of the Fourth Judicial District with office at Las Vegas. He represented the council district of which San Miguel county is a part in the legislative assemblies of the years of 1901 and 1903.

On September 25, 1895, Mr. Spiess married Miss Ruby Lynch. Mr. Spiess vacated the office of district attorney by expiration of term of office, in March 1903. He is a member of the Board of Regents of the Normal University of New Mexico at Las Vegas; and practices his profession in that city.

CARL A. DALIES, Vice-President.

CARL A. DALIES.

Carl A. Dalies, is the son of a German Lutheran clergyman. His birth place is Menominee Falls, Wisconsin, a suburban town of Milwaukee, Wisconsin. The date of his birth was December 5, 1875. In early childhood, he accompanied his parents to Racine, Wisconsin, where they resided one year. In 1877 with his parents he removed to Ripon, Wisconsin, where his father, the Rev. Carl Dalies, is still pastor of the Evangelical Lutheran church. The object of this sketch received his early education in the schools of Ripon and at the age of eighteen came to Belen, Valencia County, where he entered the employ of his uncle, John Becker, who there conducted a large mercantile establishment, now owned by the Becker Mercantile Company and with which Mr. Dalies is at present connected. In 1900 he was elected a member of the thirty-fourth legislative assembly without opposition. He served a second term in the thirty-fifth assembly and was again elected on November 8, 1904, to represent Valencia county in the lower house of the 36th assembly. In 1903 he was appointed by Governor Otero a member of the Territorial Irrigation Commission, and of the Board of Managers for New Mexico of the Louisiana Purchase Exposition, of which board he was elected vice president.

W. B. WALTON, Secretary.

WILLIAM B. WALTON.

William B. Walton was born in Altoona, Pa., January 23rd, 1871. His father Louis Walton, who died in 1890, was of staunch old Quaker stock. His mother, who is still living at Altoona, is a daughter of William Bell, deceased, one of the pioneers of the central Pennsylvania section. Mr. Walton spent his early life in Altoona, completing his common school education in that city. Later he attended the South Jersey Institute at Bridgeton, N. J., graduating therefrom in 1891. He then resumed newspaper work, in which he had previously engaged, for a brief time in Altoona, and in August, 1891, came to Deming, New Mexico, where he entered the law office of S. M. Ashenfelter, Esq., and under whom he studied, being admitted to practice in 1892. Shortly after coming to the territory, Mr. Walton became connected with the Deming Headlight, purchasing the plant in 1893, and conducting the same until 1898, at which time he disposed of the property and became the proprietor of the Silver City Independent, which he still owns. In 1895, he was appointed clerk of the third judicial district court by Judge Gideon D. Bantz, moving to Silver City from Deming at that time, and held that office until the change of administration in 1898. In 1900, Mr. Walton was elected on the Democratic ticket as the representative from the thirteenth district, and served as a member of the thirty-fourth legislative assembly. In 1902, he was elected to the office of probate clerk of Grant county, and was re-elected in 1904. Mr. Walton was named by Governor Otero as a member of the Board of Louisiana Purchase Exposition Managers, created by the Act of March 21, 1901, and upon the abolition of said board by the succeeding legislative assembly and the creation of the present one, was re-appointed, and upon organization, was elected secretary. Mr. Walton has always been an active Democrat.

In January 1893, Mr. Walton was married to Miss Leoline Ashenfelter.

ARTHUR SELIGMAN, Treasurer.

ARTHUR SELIGMAN.

Arthur Seligman was born in Santa Fe, June 14th, 1871. He attended the public schools and the Santa Fe Academy for several years and thereafter was sent to Philadelphia where he was a pupil at the public schools as well as one of the high schools. He was a student for two years at Swarthmore College and thereafter took a complete course in bookkeeping and banking at the Pierce College of Business and Banking. He graduated from the latter college in 1888 and returned to Santa Fe, where he entered the employ of the wholesale and retail drygoods firm of Seligman Brothers with which business he has remained ever since. This was established by his uncle and father in the year 1856 and two years ago was converted into a stock company under the name of the Seligman Brothers Company of which he is the secretary and treasurer. Since he became of age, Mr. Seligman has taken an active interest in politics. He is a democrat. For ten years he has been a member of the Democratic Central Committee of Santa Fe county, and for four years served as its chairman. In November 1896 and 1898, he was the democratic candidate for member of the House of Representatives of the Legislature from this county, and was defeated by a small majority. In November 1900, he was elected a member of the Board of County Commissioners from the first district for the term of four years, and in November 1904 was re-elected to the same office for a term of two years, and on January 1st, 1905, was elected chairman of the board. He was appointed by Governor Otero and served as commissioner from New Mexico to the Pan American Exposition at Buffalo, New York, held in 1901. He was appointed by Governor Otero a member of the New Mexico Board of Managers of the Louisiana Purchase Exposition in April, 1903. He was elected treasurer of the board, a position which he still fills. For several years past he has been secretary of the Fairview Cemetery Association of the City of Santa Fe. On the 16th day of June, 1904, he was appointed by Governor Otero a member of the Irrigation Commission of New Mexico and was, immediately after appointment, elected secretary and treasurer of the same.

For the past eight years he has filled the position of Auditor of the Santa Fe Building and Loan Association and is still acting in that capacity. He is a member of the Board of Trustees of the University of New Mexico at Santa Fe, also an active member of the Board of Trade of the city of Santa Fe, and of the Historical Society of New Mexico. He was married July 4, 1896 to Mrs. Franc L. Harris at Cleveland, Ohio.

FAYETTE A. JONES, Member.

PROFESSOR FAYETTE A. JONES.

Professor Fayette A. Jones was born on August 1, 1859, on a farm twenty miles southeast of Kansas City, Missouri. His father, a school teacher and civil engineer, came from Puritan stock and his mother was a Virginian, closely related to the Lee family of Revolutionary and Civil War fame. Professor Jones received his early schooling at a common country school, where he developed an aptitude for mathematics and engineering. He remained on the farm until he was twenty-one years of age after which he secured employment in a flouring mill at Blue Springs, Missouri, working alternately as engineer, book-keeper and miller. From 1880 to 1882 he attended the Missouri State University, during his spare time being employed on the college farm receiving ten cents an hour, thus being enabled to remain at school after his father had become financially embarrassed. In 1882, he married Miss Agnes A. Cairns. The year following his marriage, Professor Jones taught a country school and engaged in surveying. From 1884 to 1889 he was city engineer of Independence, Missouri, and was also deputy surveyor of Jackson county from 1884 to 1888. From 1889 to 1892 he was a student at the Missouri State School of Mines, a portion of that time being also assistant professor of engineering and mathematics, graduating at the head of his class, taking degrees both in civil and in mining engineering. From 1892 to 1893 he was engaged in mining engineering and metallurgical work in Arizona, having a narrow escape from death at the hands of the Apache chief known as "The Kid." During the fall of 1893, Professor Jones made a preliminary railroad survey from Maxwell City, Colfax county, New Mexico, through the Cimarron canyon, over the Taos pass, through the town of Taos to the Rio Grande. In 1894 and 1895 he was engineer in charge of an expedition across the Isthmus of Tehuantepec, and from 1896 to 1898 was the government assayer in charge of foreign ores at the port of Kansas City, Missouri. During this time he acted in addition as chemist of the State Geological Survey of Missouri. It was from 1898 to 1902 that Professor Jones was president of the New Mexico School of Mines at Socorro, during the last named year being appointed field assistant of the United States Geological Survey and at present has charge of the mineral resources of New Mexico as a member of the survey, making his headquarters at Albuquerque. As a member of the New Mexico Board of Managers for the Louisiana Purchase Exposition, he gathered the mineral exhibit and compiled this volume covering the mining history and resources of the Territory.

HERBERT J. HAGERMAN, Member.

HERBERT J. HAGERMAN.

Herbert J. Hagerman was born at Milwaukee, Wisconsin, on December 15, 1871. His father, J. J. Hagerman, was at that time president of the Milwaukee Iron Company. In 1881, Mr. Hagerman accompanied his father to Europe. Upon their return they took up their residence at Colorado Springs, Colorado. In 1890, the subject of this sketch matriculated at Cornell University, Ithaca, New York, and graduated in 1904, subsequently taking a course in law at the University, being admitted to the bar of Colorado in 1896. He practiced in the offices of Hall, Preston and Babbitt until June 1898, when he went to Russia as second secretary of the United States Embassy at St. Petersburg, being appointed by the late President McKinley. During his first year in the diplomatic service, Ethan Allen Hitchcock, the present secretary of the Interior, was ambassador to Russia, and was succeeded by Charlemagne Tower of Philadelphia. Mr. Hagerman's colleague, Mr. Peirce, the first secretary of the embassy, is now assistant secretary of state. Mr. Hagerman resigned in 1901. Upon his departure he was decorated with the Order of St. Anne by the Emperor of Russia. Immediately upon his return, Mr. Hagerman took up his residence at Roswell, Chaves county, to cooperate with his father in the management of the South Spring Ranch and Cattle Company, the Felix Irrigation Company and other interests. He was an alternate from New Mexico to the Republican National Convention at Chicago in 1904. He was appointed by Governor Otero in March, 1903, a member of the New Mexico Board of Managers of the Louisiana Purchase Exposition.

EUSEBIO CHACON, Member.

EUSEBIO CHACON.

Eusebio Chacon was born at Peñasco, Taos County, in 1870. During his childhood his parents moved to Trinidad, Colorado, to which point he accompanied them. There he attended the public schools and later rounded out his education at the College of the Jesuits at Las Vegas. In 1887 he entered the law department of the University of Notre Dame, Indiana, and graduated in the summer of 1889. After graduation, Mr. Chacon accepted a position at Durango, Mexico, but was compelled to return to the United States by failing health. He was admitted to the Bar of the state of Colorado and opened a law office at Trinidad, Colorado. He held the position of translator for the United States Court of Private Land Claims from July, 1891, until it went out of existence on the 30th of June, 1904, by limitation of law. In the summer of 1904 he removed to Trinidad, Colorado, where he is now a practicing attorney. Mr. Chacon is married to a daughter of Senator Casimiro Barela of Colorado.

www.ingramcontent.com/pod-product-compliance
Lightning Source LLC
Chambersburg PA
CBHW060318200326
41519CB00011BA/1760